# THE Consumer's Guide to HUSBAND MATERIAL

# *THE* Consumer's Guide *to* HUSBAND MATERIAL

Rita Justice, Ph.D.

Kimberly Nelson, D.J.M.

PeAK Press

HOUSTON

**PeAk Press**

2402 Westgate, Suite 200
Houston, Texas 77019-6608
Telephone: 713-528-6680 Fax: 713-528-6577
peakbook@flash.net

Copyright 1999 Rita Justice and Kimberly Nelson

All rights reserved. No part of this book may be reproduced in any form or by any means, electronic or mechanical, including photocopying, recording, or by any information storage and retrieval system, without permission in writing from the publisher.

Grateful acknowledgment is made to the following for permission to reprint previously published material: Frances Vaughan, excerpts from *Shadows of the Sacred*; the Boy Scouts of America, the Scout Oath and the Scout Law; "Susan Cheng and Randall te Ve" story, copyright © 1998 by *The New York Times*.

Cover and text design: Nita Ybarra

FIRST EDITION
Printed in the United States of America
Printed on recycled paper

Library of Congress Cataloging-in-Publication Data
Justice, Rita.
The consumer's guide to husband material / Rita Justice, Kimberly Nelson. — 1st ed.
p. cm.
ISBN 0-9605376-5-1
1. Mate selection.  2. Men—Psychology.  3. Typology (Psychology)
I. Nelson, Kimberly, 1959–.  II. Title.
HQ801.J88     1999
646.7'7—dc21                              99-19570
CIP

9 8 7 6 5 4 3 2 1

*To Blair and Jeff,*
*who taught us how good husbands can be.*

**OTHER BOOKS BY RITA JUSTICE, PH.D.**

*ALIVE AND WELL:*
*A Workbook for Recovering Your Body*

*THE BROKEN TABOO:*
*Sex in the Family*
(with Blair Justice, Ph.D.)

*THE ABUSING FAMILY*
(with Blair Justice, Ph.D.)

CONTENTS

**ACKNOWLEDGMENTS**
*xi*

**INTRODUCTION**
*1*

*One*

**WHAT IS HUSBAND MATERIAL?**
*THE ESSENTIAL FEATURES OF HUSBAND MATERIAL*
*13*

*Two*

**THE BAD HUSBAND**
*35*

*Three*

**DO YOU *REALLY* WANT A HUSBAND?**
*53*

CONTENTS

*Four*

## HOW TO CHOOSE THE RIGHT BRAND HUSBAND FOR YOU
### WHAT AM I LOOKING FOR IN A HUSBAND?
### WHAT-I-HAVE-TO-HAVE-IN-A-HUSBAND JOB DESCRIPTION
**69**

*Five*

## BRANDS OF HUSBAND MATERIAL
**91**

THE ADDICT

THE BIG DADDY

THE BUBBA

THE BULLY

THE COUCH POTATO

THE EAGLE SCOUT

THE JOCK

THE MACHO MAN

THE MAMA'S BOY

THE NARCISSIST

THE OUTDOORSMAN

THE PETER PAN

THE POET

THE POWER SEEKER

THE ROMANTIC

THE WORKAHOLIC

*Six*

## RATING THE BRANDS OF HUSBANDS
*137*

*Seven*

## LIVING WITH YOUR CHOICE
*171*

## INDEX
*185*

## RESOURCES
*191*

## ABOUT THE AUTHORS
*192*

## ACKNOWLEDGMENTS

WE NEVER COULD HAVE written this book without knowing a lot about men—and the women who shared at least part of their lives with them. To all Rita's clients and all the other people who have taught us about husband material, we bow with appreciation.

Many people have given encouraging laughs and words to us on the road to creating this book. To all those who said, in their own ways, "Hurry up and finish it! We need it now!" we say thank you. Most especially, we thank the members of the first focus group, who waded through an unedited manuscript version to give us their wisdom on making this book better. So to Gabby Byrum, Renata Golden, Bridget Jensen, Grace Pierce, Wilma Schindler, and Lisa Stelly, we hereby award each of you an honorary D.J.M. degree (Discriminating Judge of Men).

## ACKNOWLEDGMENTS

Gail Donohoe Storey, author of *The Lord's Motel* and *God's Country Club,* wonderfully funny and wise novels about men, women, and love, did us the honor of editing our manuscript with the intent of bringing it up to bestseller quality and beyond the reach of caviling critics. If *The Consumer's Guide to Husband Material* succeeds in either respect, Gail deserves her share of the laurels.

No matter who's getting the glory of the checkered flag, every winning race team has an excellent pit crew. Holly Smetek has single-handedly been that for us, pumping up flat tires, refilling empty gas tanks, and waving us back on the track again.

Finally, we owe huge gold trophys to Blair Justice and Jeff Nelson, our husbands. They never failed to think *The Consumer's Guide to Husband Material* was a great idea. It's easy to see why we rank them highly as husband material.

## INTRODUCTION

## WHY A CONSUMER'S GUIDE TO HUSBAND MATERIAL?

TAKING A HUSBAND is like buying a car without a warranty. You can't return it. Any breakdowns are your problem. If you choose badly and repairs become too costly, the best you can hope to do is abandon your choice by the roadside and hope to get a better deal next time. Having a husband can be one of heaven's greatest blessings or hell's worst curses. There are great deals when it comes to husbands, and there are big lemons.

The good news is that you don't have to wait until you have made your selection to know what you're getting. *The Consumer's Guide to Husband Material* eliminates the guesswork. You no longer have to consult friends, roommates, coworkers, and family to gather opinions about the guy you are considering selecting as your husband. Others' opinions are certain to be biased anyway. After all, family and friends have a vested interest in the kind of mate you select and their ratings are influenced by their own experience in husband choosing. Tom and Ray Magliozzi, Click and Clack, the Tappet Brothers of PBS Radio's "Car Talk," tell a story about their father's favorite car. "He has a Lincoln Town Car," they explain, "and he swears it's the best car he's ever had. It never breaks down. Of course, he only drives it four hundred miles a year! The rest of the cars he had he drove like the rest of us drive our cars, and they had to be repaired sometimes." So don't count on other people's opinion of what you need in a husband. Their repair history in husbands colors their advice. You know best what you need, and our guide will tell you what you are getting. Of course, this guide has its biases, too. But what we write in this book is based not only on our experience and interpretation but also on evidence. Most importantly, the *The Consumer's Guide to Husband*

INTRODUCTION: WHY A CONSUMER'S GUIDE TO HUSBAND MATERIAL?

*Material* is unbiased in that we don't care *whom* you marry. We just want to help you choose the right brand man for you.

It's shocking and pitiful that up until now there has been no central guide a woman could consult for evaluating the pros and cons of different brands of husbands. Numerous guides to selecting cars, computers, dishwashers, bicycles are available. But where's the guide that rates and ranks the choices in husbands? At last, it's here! Women need to be more informed about what kind of husband material different kind of men make. Too many women still feel and act as if they are the chosen instead of the chooser. You wouldn't think of buying a car that just happened to roll in your driveway. No, you'd think about what you ideally would want and then consider the options given your requirements and resources. You can do the same thing in selecting husband material. Women are the best shoppers in the world, but, why when it comes to men, do they suspend that consumer expertise? Too much emotion and too little thinking go into the selection process. While it's true that women are more empowered in their relationship with men than in generations past, when it comes to mate selection there is a "settling for" passivity, a kind of "we'll see what happens" attitude that can be dangerous. We aim to counter that attitude with this consumer's guide.

*The Consumer's Guide to Husband Material* will help you think through your stipulations and see which brand of husband will best fit them. Even if you know you are researching and evaluating men as to marriageable material, we want you to be informed in your choice and to know what you are getting—for better or for worse. You can't know for certain how any person will be in the future any more than you can know what kind of car you are getting until you make it your own. But a test drive will give you the feel of how it operates, and our ratings will tell you others' experience with that model. This consumer's guide will give you a good idea of what to expect from different kinds of men on the twelve features we think are essential to good husband material.

As we all know, men differ as much as automobiles, and like cars, each kind of man has his own characteristics, strengths, and weaknesses in areas women want in husbands. There are brands of men we will discuss who don't have much going for them as far as being husband material. It doesn't mean you shouldn't choose one of them, but our rating guide will help you know what you are likely to get if you select one of these models. Your experience with any one man may differ from our ratings, but our evaluations are based on evidence gathered from hundreds of experienced consumers of husband material. It's essential that you be thoughtful and well-informed because the brand of husband you choose is the brand you will get. If you select a Corvette, you will get a Corvette, not a Suburban. YOU CANNOT CHANGE THE TYPE OF HUSBAND MATERIAL YOU GET. Sure, people do change in relationships, hopefully for the better. But at our core we remain who we are. So it's very important that you select wisely. The marriage vows include the promise "until death do us part," and that's what most of us hope will be the case. We wish that for you too, so much so that we have written this guide to maximize the odds that you choose someone you can live with until the end.

Don't worry about our *Consumer's Guide* being an unromantic way to pick a mate. Romance is great, but it's a very unsmart way to pick a husband. We've all known women who were swept off their feet, head-over-heels and "crazy in love" when they rushed down the aisle. That's like buying a car when you are drunk. You might get lucky and pick the perfect model, but the odds are great that you will make a very big mistake. It's a great feeling, being in love, but we want you to be able to have that feeling last a lifetime. It's not unromantic to analyze the man you are considering for a husband. Romance takes planning and preparation: buying the candles, wine, and flowers; picking the music, the place, and the clothes. Someone's got to do the planning for your life, and it might as well be you.

At some point, wisely selecting a husband requires the detachment of a *Consumer Reports* or a *Road and Track* test driver. You have to look with the eyes of an objective observer, which means you have to be free of self-conning. None of us can do that perfectly, but all of us can do it more than we do, and when it comes to this life-altering decision of

INTRODUCTION: WHY A CONSUMER'S GUIDE TO HUSBAND MATERIAL?

picking a brand man for your husband, you had best look at him and the relationship with the inquiring eyes of a skillful mechanic. See what's there, so that you can see what has to be fixed and whether or not you are getting what you think you are.

This book isn't just about men. It's about you. For you to know which man has good husband potential for you, you have to know what you want. We'll tell you in Chapter 1 what makes up a good husband, but a good husband might not be what you even want or expect to have. Read Chapter 1 first so that you have some standard for comparison with the guy or guys you are considering. Then when you look at the rating chart of husband material brands, you'll have a better idea of what features are strong or weak in each brand. This chapter also describes the essential features of husband material.

It's as important for you to know what a bad husband is as to know what's a good one. We spell out who qualifies for bad husband in Chapter 2. Whether or not to take a husband at all deserves some serious consideration. Chapter 3 will help you answer the question, "Do you *really* want a husband?" There are definitely some hazards and drawbacks to having one. If you are just shopping with no intention of buying, that's fine with us. But be clear about your intention. Sometimes all the lamenting about the shortage of good men is actually a good defense that protects you against actually getting one. Shopping for anything can be fruitless if you don't first define your goal. You need to know what you are looking for and, more importantly, what you expect to find. The process of knowing yourself well enough to decide who's the right kind of husband material for you takes some thinking.

In Chapter 4, we'll ask you questions that will lead you through a course of self-examination about your conscious and unconscious expectations of a husband and help you compare those expectations with your job description for a husband. Chapter 5 discusses brands of husbands and the psychological/behavioral make-up of each model—how they think and feel and how they behave. The report card on all the brands is in

## INTRODUCTION: WHY A CONSUMER'S GUIDE TO HUSBAND MATERIAL?

Chapter 6. It rates models of men according to the features women look for in husbands. We may have left out some of the performance areas that matter in your book, but we have rated the brands of husbands in areas that we think are most important for satisfying, enduring marriages.

Finally, we end in Chapter 7 with some thoughts on how to live with your choice of brand husband once you've got one. This is not a marriage manual. There are hundreds of books on marriage. But we want to share some thoughts on how to understand, learn from, and live well with the husband material you select.

By now you might well be asking what qualifies Rita Justice and Kimberly Nelson to rate and rank husband material. It is indeed true that every woman is something of an expert on husbands. After all, we have all seen plenty of them, and some of us have even had several. *The Consumer's Guide to Husband Material* is not based merely on our personal experience. Rita is a clinical psychologist who has worked with hundreds of men and women in the course of doing psychotherapy for nearly three decades. Every brand of husband has shown up many times, and, in the course of marriage counseling, their strengths and shortcomings became apparent. The characteristics of good and bad husbands presented here are distilled from these years of her experience. Just as *Consumer Reports* compiles the opinions and experience of thousands of consumers into a rating of different products, we have done that for you with the experience of hundreds of women who have consulted Rita.

Kimberly, in addition to being vice-president of marketing for GDH International, has a qualification that makes her uniquely suited to co-author this book. She has a D.J.M.* degree (*Discriminating Judge of Men). It takes years of training on the job to get this certification and experience with both bad and good husbands (in that order). To be certified as a D.J.M., you have to know what you need in a husband, discern what brand of husband material best fits your needs, and take that man as your husband. She has a no-nonsense eye for the assets and liabilities in all kinds of men. We hope many, many of you will qualify for the D.J.M. certification after using this guide to help yourself in the selection process.

Just to make sure we hadn't missed any brands or rated them wrongly, we asked a focus group of women to review the manuscript and add their collective and individual wisdom. These women represented every age from the twenties to the fifties. They represented every marital status: never married, divorced, happily married, oft-married, remarried, and widowed. This group of women served as our advisory group of experts, and we have incorporated their advice and experience.

Finally, we consider ourselves qualified to offer this consumer's guide for one other reason: we both have good husbands. It has been said that you shouldn't advertise your man, but we could hardly hold out hope for you if we had not succeeded in finding good husbands using the criteria and rating guide we offer you here. Our success in having good husbands is not due to beginner's luck. We both were definitely uninformed consumers in the past and chose poorly before picking our present husbands. But wisdom isn't about not making mistakes. It is about learning lessons from whatever we experience. Oliver's Law says, "Experience is something you don't get until just after you need it." We have learned from our own and other women's mistakes, false choices, and experience. *The Consumer's Guide to Husband Material* shares with you that wisdom gleaned from hundreds of women's wrong and right choices so that you don't have to rely only on your own experience in making your decision. You still have to make your own choice, to be sure, but we believe our book will help you do so with more confidence. So proceed on now with the assurance that you too can be an informed consumer of husband material!

## One

## WHAT IS HUSBAND MATERIAL?

MARTIN ABZUG DIED IN 1986, and Bella talked to him every day after his death. Pioneer feminist and former Congresswoman Bella Abzug said of Martin, her husband of forty-two years, "There isn't a thing that happens that doesn't remind me of him. I talk to him. I dream of him. I dreamt not long ago that we were dancing. He was a great dancer....I still hear him saying things....It was like having a warm fire: wherever I went over the years of struggle, there was always one place where the hearth was—Martin. I was enveloped by the warm fire of him." (*Ms.*, July/August 1990, pp. 94–96) Bella moved on in her life, enjoyed good relationships with many people—male and female—and, until her death in 1998, was nourished by the memory of Martin.

Martin Abzug was obviously good husband material. Long after his death, his absence was painful in Bella's life, but she found growing comfort from daily talks with him. He might not have been the kind of husband you need, but he had the traits that make up a good husband, traits that are important in the make-up of the fabric of quality husband material. What are those traits?

The *Oxford English Dictionary* defines a husband as "a man joined to a woman by marriage; one who manages his household, or affairs or business in general, well or ill, profitably or wastefully, etc." (*OED*, vol. VII, pp.510–511) Husband material is "a person who has qualities suitable for" being a good husband. The dictionary doesn't give us any clue as to what those qualities are, but, fortunately for you, we've put together a synthesis of women's collective wisdom about those qualities. Automobiles have to have certain basic features to qualify as a usable means of transportation. A car has to have an engine that runs. It has to be able to start and stop as necessary, to be steered to a desired destination, to be able to carry passengers. A means of securing it from easy theft and to protect the people inside from unwanted intrusion is obviously desirable, but it is still an automobile even if the locks on the doors don't work. The style of car is only a matter of preference. A Ford is just as much an automobile as a Toyota. The same is true of husband material. Certain basics are necessary for a man to be considered "suitable." The brands of husbands differ in what they come equipped with, but the basic components remain the same for any good husband.

    A client of Rita's was the original inspiration for this book. We'll call him Albert. Albert was handsome, rich, had a job, and his wife was beautiful. "You're not husband material, Albert, and I don't think you ever will be," Rita told him at the end of one session. He wasn't convinced. Why should he be? He had a wife. The fact that she was miserable because of his cocaine addiction, infidelity, and dependence on his wealthy parents was her problem, not his, he thought. The point is that some men, like some cars, don't have the basics to make them serviceable as husbands. They are perfectly capable of getting married but not of being good husbands. A man's willingness to walk down the aisle does not qualify him as husband material.

## THE ESSENTIAL FEATURES OF HUSBAND MATERIAL

So what are those qualities essential to husband material? There are twelve. We've listed the traits alphabetically, not in order of importance, because they are all important. The fundamental traits of husband material are:

1. COMMITMENT
2. COMMUNICATION
3. FAITHFULNESS
4. FRIENDSHIP
5. HONESTY
6. HUMOR
7. INTIMACY
8. MATURITY
9. RELIABILITY
10. RESPONSIBILITY
11. ROMANCE
12. SEX

Be assured that we know no one man is perfectly endowed with all of these characteristics, but if any of them are totally lacking he won't be solid husband material any more than an automobile without an engine is really an automobile. The choice is yours. Even though some qualities that we list as "essential" are missing, you may decide the man you want has enough good traits that you'll opt for him. But we want you to know, as the comedian Flip Wilson used to say, "What you see is what you get!" If you see that many of the essential features we list here are missing, you will get a husband but he won't be a good one. He might be one you can live with, but there will be some problems. If the man you are interested in gets failing marks in any of the categories, the problems are likely to be very serious. A passing grade means he'll do OK and may improve over time. If it's all right with you that he stays a "C" student in some area, it's all right with us. Few guys are going to rate all "A's," and it isn't necessary to be a straight-A performer in every category in order to earn the ranking of a good husband and for you to love him very much. But how good a husband can a man be if he has limited capacity to love, has poor commitment to the relationship, is weak on honesty and integrity, is disloyal and untrustworthy, doesn't respect you, is humorless, immature, and unspiritual? Not good, we promise. People are more likely to change in some of the categories than in others. Most people, men and women, mature as they experience more of life. Respect grows as people come to value each other's contributions over time. Spirituality comes late to many, regardless of sex. Life does change people. But we have to evaluate what and who is before us right now. Your guy had better score a passing grade or better in each of the essential features or you are headed for big bumps on the road of marital life ahead.

You may wonder why we haven't listed "intelligence" as an essential feature in husband material. We haven't because it's not one. The brands vary widely within themselves on this quality, and an intelligent man is no more (or less) qualified. Smart men can be lousy husbands. Einstein is an example. If you think it was fun being Mrs. Einstein read one of his biographies. So we think intelligence is an optional feature, like antilock brakes or automatic transmission. His being intelligent might make your life a little easier, but it's no guarantee. Emotional intelligence, on the other hand, is important. (See Chapter 2.) Emotional intelligence is reflected in many of the essential features, and it's his grades on the essential features that count.

We'll now discuss these essential features that make up husband material. Think of this as a list of manufacturer's standard features.

## COMMITMENT

Commitment is like a warranty but not quite the same thing. Commitment is a man's willingness to take you down whatever road the two of you need to travel. The promises made in the marriage ceremony are declarations of commitment: to love, comfort, honor, *keep,* in sickness and in health, forsaking all others, being faithful, as long as you both shall live. Commitment is his intention to keep those vows. It encompasses faithfulness, as does the marriage vow, but it's more. Commitment is the dedication that makes faithfulness possible.

A man with this essential feature is one who willingly relinquishes his future options to rent, lease, or borrow another when he says, "I do."

## COMMUNICATION

With some cars, communication is difficult. Simple actions, like adjusting the side mirrors or turning on the bright lights, are exercises in frustration. Whoever designed the car doesn't seem to have any clear idea of how people do things. Some guys are like that, too. Just getting them to understand what you want takes much more effort than it seems should be necessary. While it's been said that men and women come from different planets and therefore speak different languages, communication is mostly a matter of wanting to understand the other person. A good husband is the kind of man who genuinely wants to know what you want him to know and wants you to know what he wants you to understand. In short, he wants to understand you and to be understood by you. Even if it's difficult and requires effort, just as communicating in a foreign language requires effort, a willingness to communicate is a necessary trait in any good husband.

### FAITHFULNESS

Faithfulness in husband material means that he won't go rolling off just because you forget to put on the emergency brake. He stays parked in your driveway because that's where he wants to be. Some men act as if they are still on the dealer's lot long after you have taken them home and have the title papers in your lockbox. Faithfulness means he believes and acts as if he belongs with you. We are talking sexual, psychological, and emotional loyalty. Loyalty is a synonym for faithfulness. Loyalty means putting you at the head of the line in his mind. You count more than others do. Good husband material is someone who wears an invisible bumper sticker that says "Charity begins at home." It is an implicit or explicit statement that he'll be yours no matter what. You may not be able to count on him to remember to pick up the cleaning, but you can rest assured that he won't forget he's yours "until death do us part."

### FRIENDSHIP

A Kanji is the Chinese symbol that represents two friends holding hands. It represents friendship. Here's what it looks like.

$$友情$$

Kanji is a symbol to look for in husband material. A guy who isn't your friend isn't going to make good husband material. He doesn't have to be your best friend, but he needs to be someone you at least count in your list of friends. By friendship, we mean a relationship in which two people like each other, enjoy being together, and care about one another. Many of the traits we list as essential for good husband material contribute to friendship—communication, intimacy, honesty, commitment—but friendship itself needs to be part of the relationship between you and the man you are considering as a possible husband. It's astonishing that women would consider marrying anyone they aren't really friends with, but some do. It's like buying a car you really don't like. That's OK with a car if you have to make that choice. You never have to make that choice with a husband.

## HONESTY

Honesty is more about the car salesman than the car. Here the issue is truth in packaging. A man that is good husband material isn't 1) afraid of your knowing who he really is; and 2) afraid of your knowing what he's really doing. He doesn't represent himself as a Mercedes when, in fact, he's a Toyota Corolla. Of course, this requires reciprocal honesty on your part. If you are hiding the truth about who you really are, you can hardly expect "truth in advertising" from the salesman who's trying to sell you himself.

Honesty involves integrity. We like how the dictionary defines integrity. Integrity is "the condition of having no part or element taken away or wanting: undivided or unbroken state, material wholeness, completeness, entirety." In a husband, this means that all you need is there. A Volkswagen has integrity as a Volkswagen. It does what it is expected to do. A Lexus has to have integrity as a

Lexus. The two are not the same. You need to be clear about what constitutes completeness for you. What matters is not whether a man is complex or simple, driven or laid-back. Is he who he says he is and represents himself to be? Integrity means he is honest about himself—to you, to himself, and to everyone else.

*HUMOR*

Life is too hard not to be able to laugh with the person who is going to see you through potholes and blowouts as well as drive with you down superhighways. This feature, humor, may seem optional for a husband, but laughter brings a kind of bonding that is different from any other. It is a kind of affirmation that you both see the world in the same way, at least at the moment you both crack up laughing. It's not necessary for a man to be Steve Martin or Robin Williams. His job as a husband isn't to entertain you. It's to be able to share with you the delicious moment of letting go in laughter. We know a woman whose husband bought her a Mercedes 500 SEL, but he never laughed with her. She had an affair with a man who did.

## INTIMACY

Intimacy is a no-holds-barred relationship. It's a way of relating from your deepest nature, from the essential you. Some people, men and women, just can't do it. It's too scary for them to drop the persona or the pretense of who they want you to think they are. You can't expect any more openness in a husband than you can risk for yourself. If you choose a brand of man who is comfortable with who he is and willing to be open about his "deepest nature," it makes it much easier to take the risk of being open yourself.

Intimacy has much to do with the capacity to love. The more you genuinely love yourself the more you can love someone else and the easier it is to be your true self with those you love. Intimacy is a natural outgrowth of love. When we love, we feel connected to the one we love. We feel the closeness and oneness that poets and love songs tell about. So intimacy with a loved one is directly proportional to the capacity to love. He can only love you as much as he can love anyone. This is engine size we're talking about. Bella described Martin as having a "generosity of heart." Blair Justice, in his book *A Different Kind of Health: Finding Well-Being Despite Illness,* (Peak Press, 1998), explains about hearts:

> Hostility, cynicism, pessimism, depression, and anxiety are all well-documented feelings and attitudes that undermine the health of the heart and the cardiovascular system. There is truth, then, in the poetic notion that the heart is the seat of love and compassion as well as in the scientific view that it is a pumping mechanism. (p.5)

Some guys have hearts like the battery in an electric car or the engine of a 4-cylinder Yugo. They can do the job but not with much power or enthusiasm. For others, the seat of love seems to have gotten stuck in their gonads. The capacity to love is, by and large, determined by the manufacturer. If a guy didn't have much love and nurturance as a baby and little kid, his capacity to love is likely limited to the amount he experienced. He can't love you more than he knows how to love.

Sometimes things happen along the way that make a potentially powerful engine run poorly. In those cases, the capacity to love and to be intimate is ample but substantial rebuilding may be necessary. The rebuilding is *his* job, not yours, but your love can serve as a mechanic's garage where he can do the work. The bottom line is that if he can't love you, he can't be emotionally intimate with you over the long haul, and there's nothing you can do that will fix that.

### *MATURITY*

Maturity is "off road/on road" capability, performance under all kinds of road conditions. Mercedes-Daimler put out a new little model, the ML320, which looks like a cross between a van and a traditional off-roader, but it's supposed to provide car-like highway handling never before attained in a 4×4. Orders were pouring in. Then someone decided to put the new model to the "moose test." That's a test that determines how the car would respond if the driver had to swerve to avoid a moose. The results of the moose test on the new Mercedes were that the moose would survive, but the driver and passengers would be killed because the car flipped over. There are plenty of moose on any marital road. Maturity in a husband means he can keep all wheels on the ground no matter what kind of maneuver is required.

One mark of maturity is fairness. The mature man doesn't have to win every time. He knows that taking turns is the right thing. Automobiles don't know about taking turns, but good husbands have to have this figured out. The dictionary (*OED,* vi, p. 675) says fairness is "equitableness, fair dealing, honesty, impartiality, uprightness." In husband material, this means that how the game is played counts as much as who wins. Men with maturity are fair with their wives, in bad times and good, about everything from sex to money.

### *RELIABILITY*

Reliability means trustworthiness. You can count on him. If you fall, you know with certainty he'll be there to catch you. The catch may not be graceful or as sure as you might have liked, but you know he'll be there when you really need him. It's like the capacity of a car to start up even in the most bitter weather. You can have confidence in him, like you do in a car that carries you safely to your destination no matter what the weather conditions. Rita always thanks her car when it starts up after being left at the airport parking garage for several days. She appreciates the car's goodwill in being operative under all conditions. Reliability in husband material means he will do what he promises or agrees to do. No tricks. No games. No excuses. He can be counted on. You can see why a guy with this feature makes having a husband a whole lot easier.

### *RESPONSIBILITY*

Responsibility literally means being responsive. It requires agreement on what each partner will be responsive to and about in and out of the marriage. Responsibility in husband material is a measure of how heavy a load he can easily and willingly carry. It's his "weight load bearing capacity." Most men these days know they are expected to share more of the responsibility for the maintenance of family and marriage than their fathers did, but some of them still do this with what can euphemistically be called reluctance. A guy who shirks responsibility whines when you ask him to do something or, when you ask him to help, follows you around the house asking you what you want him to do and how you want him to do it. He makes it so much work for you to get him to help that you end up deciding it's easier to do it yourself.

Responsibility and maturity go hand in hand. The brand of husband who shows maturity is also one who knows that part of being a grown-up is to do his own share of the work and, if necessary, his wife's share, too. Most everyone's life is complicated these days. Women are doing more, in less time, and with less help than ever before. You want a car that makes the trip easier, not one that you have to push out of the driveway. A man who willingly assumes responsibility makes the marital journey much less tiring.

## ROMANCE

We list romance as an essential feature in husband material because it's important in maintaining the specialness of the marriage. It's like riding in a car that's just been detailed. It's still the same car, but it feels more special. A man's willingness to go to the trouble of being romantic occasionally reflects his understanding that his relationship to you is special and needs to be acknowledged as such. He doesn't have to be as romantic in the marriage as he was in courtship. Part of courtship is playing the roles of two people who are romantically in love. But he has to be willing to do some special things that most people would consider romantic on appropriate occasions, like anniversaries, Valentine's Day, and sometimes for no reason at all. Romance is the seasoning that spices up the ordinary of a marriage. Of course, too much seasoning can ruin a meal. With some brands of men, romance is about all you'll get. No potatoes. Just gravy. You are right to want some romance. Just don't count on it to be a balanced diet. A brand man who rates high on romance but low on other essential features can leave you feeling like you got a tank of bad gasoline.

### SEX

Sex is the oil in the engine of a marriage. It has to be sufficient quality to match the engine's capacity, but it doesn't have to be the highest grade on the market. Certainly there comes a time in a marriage, if both people are fortunate enough to live together until a very old age, when the engine runs on some other kind of lubricant, like love, caring, and shared history. But you're not at that stage. When you are shopping for husband material, sex is important. It doesn't have to be the greatest sex ever written, told, or lied about, but it has to be good enough for you. The match between your expectations and those of the brand of guy you are considering needs to be carefully considered and discussed. And, by all means, take a test drive! You cannot possibly know how the ride will be until you take it on the road. If your values proscribe intercourse before marriage, no problem. We know virgins who have fantastic safe and responsible sex lives! A test drive doesn't mean you have committed yourself. It's just a great and, hopefully, fun way to find out to your satisfaction if a guy has this essential feature of sexual compatibility.

So that's the list of essential features for good husband material. In the next chapter, "The Bad Husband," we'll discuss what disqualifies a man from being husband material. We'll give warning signs that will let you know that the brand you are considering is a potential lemon—or worse.

## Two

## THE BAD HUSBAND

EVEN THE WORST HUSBAND has some value. Dr. Bernie Siegel, the physician and well-known author of *Love, Medicine and Miracles,* says his wife Bobbie tells him, "Never consider yourself a failure. You can always serve as a bad example." No matter how unsuited a man is for husband material, he can serve as a model of what kinds of breakdowns you can expect if you chose him as your brand.

There are five reasons that earn a guy the classification of "bad husband."

1. He is a man who is incapable of caring how you or anyone else feels.
2. He is a psychologically abusive man.
3. He is physically abusive.
4. He is a man who doesn't see you for the person you really are.
5. He's the wrong man for who you need for a husband.

## DISQUALIFICATION #1:
*A man who is incapable of caring how you or anyone else feels.*

Some men have the sad misfortune of never developing emotional empathy. Whether the cause is childhood abuse or a bad gene pool doesn't really matter. They are impaired emotionally and morally and have such major design flaws that they are simply unworthy of having a wife. That may sound like a harsh sentencing for any man, but, sad to say, it is true of some guys. Every job has its requirements, including the job of husband. Some men cannot meet the basic minimum requirements, like caring how you feel, for a starter. We are not saying these men have no value and are not souls worthy of love and compassion. They are that. They just shouldn't be considered as husband material. There are psychiatric names for this brand of man: sociopath, psychopath, character disorder. You have names for him, too: con man, creep, rat. Unfortunately, you may only start calling him those names after you have been taken in.

Like the spider, some of these guys spin a beautiful web, but to step into it is dangerous, psychologically for sure, possibly financially, and even physically. Most men can be unfeeling at times, as women can be, but they nonetheless have decent values and behave morally. That's not who we're talking about here. Feelings of caring and compassion can be reawakened if they weren't extinguished by terrible parenting or some other soul murder. But you cannot rebuild the psychological engine of a man's psyche. Just pray for him and pay for your own drink. Here are some warning signs to help you spot a man we would disqualify as husband material for reason #1—lacking feelings.

WARNING SIGNS:
- His basic life position is I'm OK/You're Not (meaning everyone else but him). This warning sign is easy enough to spot by listening even a few minutes to how he talks about others. Is he quick to judge others as losers, fools, or idiots? Does he make it a point to talk about how smart, clever, and right he is? Does he take pride in out-

smarting others? Does he belittle other people's or your accomplishments? If the answer to most of these questions is "yes," his basic existential position is I+/U–.

- His basic life position is I'm Not OK/You're Not OK. Actually, this is the true psychological position this guy lives in because "I'm OK" is a con, like the rest of his life. He doesn't really think he or anyone else is OK, but he has such a good line, he even convinces himself sometimes. These are some questions to ask yourself about the more obvious I'm Not OK/You're Not OK man. Does he see people as basically bad? Does he treat others and himself badly? Is his life a series of failures and made-up tales of "getting screwed" by others? Is he going nowhere? Several "yes" answers should have you reaching for your purse and car keys.
- He lacks feeling for the suffering of other beings. If seeing a dog hit by a car doesn't bring at least a wince from him, consider getting out of the car of his life at the next stop. He doesn't have to be a

man who can't swat a mosquito, but a man who cannot feel pain at the suffering of animals isn't going to feel any pain for your suffering either.

- He takes pride or pleasure in cheating others or pulling a fast one on someone. It doesn't matter how small the dishonesty, a man who likes getting away with deception is ripe for deceiving you. Does he get off on telling stories about how he tricked someone out of something? Do his values leave you feeling uneasy or queasy? Is he someone who seems to have deleted a few "Thou shalt nots…" from the list of Ten Commandments he feels obligated to follow? Be forewarned. He's not going to reinstate those commandments (like "Thou shalt not commit adultery" or "Thou shalt not bear false witness") for you. If those values weren't factory-installed, there's no way they can be wired in later, regardless of how much you are willing to pay to have what should have been there in the first place.

## DISQUALIFICATION #2:
*A psychologically abusive man.*

The guy we just described above may sound psychologically abusive, and it certainly abuses your psyche to be in a relationship with him. But the psychologically abusive man is one who has a plan. He misuses power to get you to submit and be in his control. If he succeeds, the damage to your spirit and sense of self from psychological abuse can be as serious as the consequences of physical abuse. (See Disqualification #3.)

In her book, *I Just Lost Myself: Psychological Abuse of Women in Marriage* (Praeger, Westport Connecticut, 1996), Valerie Nash Chang, an associate professor at Indiana University, lists these warning signs in a guy with the potential to be psychologically abusive.

WARNING SIGNS:

- He's a control freak, with a sense of entitlement to a position of dominance over all aspects of the relationship.

- He dominates with unrealistic perfectionistic demands.
- He requires unilateral control over finances (his and yours).
- He expects you to have sex whenever and however he wants it.
- He tries to control your social contacts by criticizing friends and family and responding to time you spend with others with sullenness, irritability, criticism, and verbal attacks.
- He's charming but phony.

Why, you might ask, would any woman even go on a second date with this man? There are two reasons. The first reason, as Chang explains, is that "Psychologically abusive relationships start out with the same high hopes that characterize other beginning relationships and often do not develop obvious problems until there is some competition for the wife's attention."(p.9). Although the man showed these needs for control and dominance early in the relationship, the woman saw them as signs of strength.

The second reason you might go out on a second date with this guy is that women who as children were neglected emotionally and lacked close emotional connections to their parents are especially vulnerable to psychological abuse from a man. In her research, Valerie Chang found that women who stayed in psychologically abusive marriages grew up feeling "like nobody," like they were "just not there." As a result, she concludes, "they had little self-confidence and craved loving attention so they were vulnerable to accepting the first love relationship available to them." (p.47)

This means that if your background and sense of self puts you in the category of women vulnerable to psychological abuse, you need to be especially alert to these warning signs. If you think you can't trust your own judgment, don't! Let a professional check him out before you do much more than take a test drive. Ideally you'd do that much with any car you were considering making your own. Rita has had many clients bring potential husband material for interviews. The women weren't always happy with her assessment but none of them regretted having the information.

## DISQUALIFICATION # 3:
### *He is physically abusive.*

Physical abuse is also psychological abuse. The intent is the same. He wants to control you, and he uses physical force to do it. He is a man who has learned that he can get what he wants through physical intimidation, and once he's succeeded, he'll keep using that means. It should make no difference to you *why* he hit you. If he hits you once, he can do it again. Don't stand around trying to understand him, and don't make any excuse for him. You have just two options that will change his behavior:

1. Tell him if he hits you or threatens to hit again, you are leaving the relationship. Then do it.
2. Leave if he ever hits you.

WARNING SIGNS:
- He has a violent temper.
- He was beaten as a child.
- He has a history of getting into fights and likes to tell "war stories" about them.
- He threatens you (or your children) physically.
- He hits you (or your children).

It makes no difference how lovable he is at other times, a man who is physically abusive is a bad husband. If he is smart enough to acknowledge that he has a problem and to enter treatment for his violent behavior, he can be rehabilitated sufficiently to be reconsidered as husband material. Until then, keep shopping. This one won't do.

### DISQUALIFICATION #4:
*A man who doesn't see you for the person you really are.*

If a man doesn't see and accept you as the authentic person you are, he cannot love you. A man who sees only an ideal image of a woman is incapable of really loving her. He will start out seeing you as the woman of his dreams only to decide later that you are the supreme wicked-bitch-whore-ballbreaker, which indeed you may become after living with this guy for awhile. Just as we challenge you to know what your expectations of a husband are, you need to be very informed about his expectations of you. While it's true that love is blind, husband selection shouldn't be.

WARNING SIGNS:

- One sure-fire way of knowing a man is more interested in his image of you than in the actual you is his investment in how you look. If he has more than a passing opinion in how you wear your hair, what clothes you choose, or the shape of your body, be cautious. Don't think you'll eventually get it right and please him. It won't work. One woman in Chang's study of psychologically abused women explained how she tried to do that: "If [ex-husband] didn't like what I was wearing, I would change. I wore my hair the way he wanted me to wear it." (p.51) It's fine to want to please the man in your life, as long as you please yourself as well.
- An automatic disqualifier is his insisting on going with you to shop for clothes. It means he either doesn't trust you or wants to be sure you come out looking like what he had in mind. He sees you as an extension of himself, like his car or his new sportscoat, and this is definitely not a good way to be seen. His car is going to get old and break down and so will you sooner or later. The bad husband will do to you just what he

does with the old car and worn-out coat: trade you in on a new model. It's lovely to have our guys care about us enough to compliment our choices in hairstyles or clothes. A man who worries about his wife's health and fitness is expressing loving concern. There's a big difference, though, between having him admire your appearance and seeing you as a remodeling project.

- He doesn't see *you*. He sees either an idealized image or a make-over project. Even if you see yourself as someone in need of a make-over and are not all that comfortable with who you are, the impetus for change needs to come from *you*, not from him. It's perfectly OK for you to have a Victoria Secret fantasy of yourself hidden beneath your smartly tailored suit, but it needs to be *your* fantasy of who you really are, not his. Drive the car of your life your way, not by following directions from Mr. Him in the backseat, but by knowing who you are and how you want to live.

If this make-over master description fits the guy you have an interest in, get over your interest right now. Get a dog. Do volunteer work. Pray. Whatever it takes, move on. Repeat to yourself every so often what we said at the beginning of this disqualification: if he is incapable of knowing you, he is incapable of loving you. Being in a marriage with a stranger is much lonelier than being alone.

### DISQUALIFICATION #5:
*The wrong man for who you need for a husband.*

The fifth reason a husband is classified as a "bad husband" is more personal than the other four disqualifiers. A man may be some other woman's bad dream, but he's just what you always hoped to find. Owning a vintage car is a delight for someone who loves to tinker with and fix old cars. It is a nightmare for a person whose interest in maintaining a car is limited to filling it up and having the oil changed every so often.

In her last year of graduate school, Rita bought a 1950 MG-TD. That was in 1969. It was a gorgeous "rebuilt" car: a silver-bodied convertible with a black leather strap securing the hood, a red leather bench seat, and, the crowning touch, a steering wheel on the right side, British style. In a long "maxi coat" and beret, with a neck scarf blowing in the wind, riding in this car was terrific fun and a great attention-getter. The fun waned quickly, though, when the car spent more time in the shop than in the driveway, and the repair bills looked like the national debt. It soon became apparent that this car was for fun, and just that. Rita couldn't count on it to get her to class on time or at all. The car went on its way, needing a new engine to replace the "rebuilt" one that wasn't really. With any luck, it's now making some vintage auto collector very happy sitting in his garage. Wherever it is, it's not Rita's problem any more.

So use this as a model for what to do with men you may love who just aren't what you need. Let them go on to some other woman for whom they are suited. Cut your losses, remember the fun, and move on. Life is too short to spend it riding in a relationship that ill suits you. Of course, you have to take some responsibility for knowing what *your* requirements are in a husband. Psychologist Frances Vaughan says in her book, *Shadows of the Sacred* (p.101):

> Major obstacles to love include the expectations one has about how others should be. When reality does not match the image, the person who chooses to ignore reality rather than give up the image pays a high price, since it implies giving up on life and choosing to live in illusion instead. On the other hand, in choosing to face things as they are, letting go of preconceptions, one may learn to love wholeheartedly.

We don't think it's a problem to have a preconceived notion of what you want. The problem is in being unwilling to test your specifications against the man you are considering as your selection.

WARNING SIGN:

- We all have friends who have conned themselves into buying into or staying in relationships that are bad for them. Their explanations are sometimes simple (or simple-minded), as in "He needs me," or elaborately convoluted (and equally stupid), as in "He's projecting his negative mother complex on me and we'll be fine once he gets through this part of his analysis." All these tales may be true, but be very certain they aren't rationalizations for buying a model that won't serve you well.

It's fine and good to promise to take him "for better or worse." Before you sign the deal, just be sure you know as fully as you can what the betters and worses are. Why would you consider a man who is completely unskilled as a husband for your future mate? It's no more necessary than buying a car that won't run. We hope you won't do either. We would like to reiterate once again a point we made in bold type in the Introduction. **You cannot change the type of husband material you get.** Like Madame Butterfly in Puccini's opera of the same name, too many women have done themselves in waiting for the man they love to come around. It's a great operatic tragedy, but is that really how you want your life to turn out?

Now that you know what makes for good and bad husband material, we encourage you to look at whether or not you really want to make a husband deal at all. It's time to ask yourself a hard question. Do you *really* want a husband? In Chapter 3, we will help you answer that question truthfully.

### Three

## DO YOU *REALLY* WANT A HUSBAND?

LUCINDA WEPT: "Why can't I have a husband and baby like every other woman? That's all I want. Just to have a husband and a baby of my own!" If you saw Lucinda, you could wonder the same thing. She's attractive, thoughtful of others, smart, hardworking. She once heard a little girl in a restaurant ask her mother, as she pointed to Lucinda, "Mommy, that lady is so pretty. Why doesn't she have a husband?" Why, indeed, especially given the fact that five men had pleaded with Lucinda to marry them? That's right, five serious offers, all from men Lucinda had been involved with at different times in her life. Clearly, something doesn't add up.

Taking a husband is basically a two-step process. First, you look over the inventory. Then you make your selection. Some women cheat themselves on the first step. The mantra is, "There are no men out there to marry." That's like bemoaning not being able to find a car to buy. The problem isn't inventory shortage. Like cars, there are millions of guys out there. The obstacle is as much an unwillingness to check out the inventory as it is finding a man you want to keep. Notwithstanding our comparison of men to automobiles, it is a much more complicated process, emotionally and socially, to decide to marry and to have it end in divorce than to junk a clunker. Understandably, like Lucinda, you may have ambivalent feelings about marrying, and well you should. Some research studies show that while married men live longer than unmarried men, unhappily married women have a shorter life expectancy than their unmarried sisters. It is definitely not an easy thing to be a married woman. So, if you find yourself singing a lamentation similar to Lucinda's, you might ask yourself a hard question: Do I *really* want a husband?

Some people, the minority, like shopping for cars. Most of us hate it. We dread the encounter with the unknown. We fear making a mistake or getting conned into a bad deal. All that angst gets agitated, and it's only a piece of machinery we're committing to own. The mate-selection process is deservedly more gut-churning. But if you *really* want a husband, you have to turn yourself into a legitimate consumer, queasiness or no. You'll never find the husband you are seeking if you aren't honestly looking. Once you are genuinely on the quest, the options appear before you.

When you decide to buy a car you quit throwing out the classified section of the Sunday paper and start looking at the ads with the intention of finding what you want. What you want and need in a husband is out there just as assuredly as is the right automobile. We wouldn't have written this *Consumer's Guide* if we didn't have confidence in inventory availability. Our intention is not to convince you to go on the quest, but that the quest has to begin with your commitment to take a husband. Once you have created that psychological readiness, the options appear in all their shapes, sizes, and brands. The next step then is to become an *informed* consumer. Lucinda and many women like her that Rita has counseled run into major obstacles in selecting husband material. The obstacles aren't an absence of good material to choose from, although many women do shop in the flea markets of husband material. Shopping in the flea market includes dating men who are not suitable husband material for you. Unsuitability may be because of married status (i.e., already has a wife), sexual orientation (he's gay), or some other clear disqualification (see Chapter 2). It's OK to browse the flea markets just to pass some time and see what's there, but it's a foolhardy way to find a quality husband. If your proclivity is to shop the garage sales and flea markets for bargains in men, we would question whether or not you are a committed consumer in the husband market. You can spend a lot of time that way, looking at trash without much probability of turning up with a treasure.

But let's go back to the major obstacles women run into in selecting husband material. The most significant obstacles women encounter in finding and selecting a husband are 1) what they expected to find and 2) expectations not matching reality. Let's talk about the first obstacle.

What you expect to find in a husband is your father. Rita's father died when she was almost four. Her expectation of a husband was someone who would take care of her and then die. As fate would have it, her husband Blair met the first criteria and, fortunately for both of them, not the second. Kimberly's expectation in her first marriage was "we'll see what happens." What happened was that she chose a man like her father, someone who couldn't commit emotionally to her or anyone else. Whether you are certain that you never want a husband like your father or you hope to find one just like him, that's what your unconscious expects. It's the yardstick we use for measuring husband material. You may end up marrying your mother, but you expect to marry your father.

Lucinda's father sexually abused her and physically abused her mother. For the most part, the men she got involved with were abusive to her, emotionally and sometimes physically, and controlled her psychologically just as her father had.

Lucinda dated nice guys, too, but she never felt any passion with them. What she expected in a husband was her father. Guys who treated her badly met that expectation. Lucinda had the wisdom to know that marriage to this kind of man would be a continuation of the nightmare she had lived through growing up. She ended up getting marriage proposals from men she feared marrying and envying women with husbands and babies.

If you had a troublesome relationship with your father or your parents' marriage was difficult, you probably bring some ambivalence to your husband quest. You might be asking yourself questions like these: "What if he turns out to be like my father?" "What if he treats me like my father treated my mother?" "How can I trust any man when my experience with men has been so bad?" Actually, those are good questions to be asking yourself because it means you are thinking *consciously* about your choice of husband. This guide can help you in that thinking. Seeing the pros and cons of each type of man will help you know whether or not you are shopping at the same dealership your mother did when she rode home with your father.

Having had a painful or abusive relationship with your father doesn't doom you to repeat history. But if you don't think about whom you are picking and why, your history could well be your future. A marriage has the potential to be the most healing relationship in your life. In fact, that's a big part of what people are looking for in a marriage—a loving relationship that will fill the holes left by childhood wounds. The greater your wounds, the more you need the relationship to be healing. It can be that, if you understand what you need from a husband and which type of man is most realistically likely to be able to meet those needs. You need not be ambivalent or despairing. You need to be an informed consumer.

The other major obstacle to finding a suitable husband is having expectations that don't match reality. What you do with that misfit creates psychological dissonance. Dissonance is that uncomfortable feeling we get when two thoughts don't fit together or when we believe one thing and do the opposite, like when we buy something we can't afford. One of those thoughts has to change for us to get comfortable again. We rationalize the purchase by telling ourselves it's OK because we really had to have it. With husband material, that kind of rationalization can be costly and even lead to emotional bankruptcy.

Frances Vaughan, psychologist and author of *Shadows of the Sacred,* tells about one of her clients who worked through failed expectations in a courageous way. (p.101–102)

> Patricia was a very successful, attractive young woman who came to see me for psychotherapy because she wanted to make her relationship with her boyfriend work better. She had been in the relationship for about two years, and she thought she wanted to marry this man. He seemed to be exactly what she had always wanted in terms of status, interests, background, education, looks, etc. Although the relationship was problematic, she was determined to make it work. First she tried very hard to change him, to make him be what she thought he should be. When this did not work, she tried making herself change to fit his ideas of what he

wanted. She thought that if she could adapt, she could make it work. Finally, she gave up. The price for holding on to the relationship was too high. Her soul felt trapped when she could not express herself honestly. She found that she was withholding grievances and pretending to be happy in order to please him, trying to hold on to what she thought was security. She felt she could not really open her heart and be all who she was in the relationship. When she finally summoned up enough courage to let go, she felt a great sense of relief. The strain of trying to make the relationship work when it was basically unsatisfying had taken its toll. Letting go of the relationship was not as difficult as giving up the dream of being happy together.

Letting go of the dream is often more painful than giving up the reality. For Patricia, giving up the dream meant confronting aloneness and a sense of failure. When she did, however, she began to feel much better. Before long she was ready to look to the future and to visualize the kind of man she wanted as a partner. Being fairly conventional in her tastes, she thought, in accordance with cultural stereotypes, that the man should be older, taller, and make more money than she did. She soon found herself dating a man that, once again, met all her specifications. She was very pleased for a few weeks, until she realized that something was lacking. Everything appeared to be going smoothly, but the spiritual connection was missing. At this stage she was feeling stronger in herself and was unwilling to waste time in a relationship that seemed constricting. Once more she was willing to let go. This time it happened more easily and more quickly, and she felt better about being on her own again. As she began to give up her conventional ideas of what she thought she wanted and to be open to following her heart, she found herself being drawn to people who were different from those she had previously known. As she gained greater emotional independence from her parents and became more self-confident, she began opening up to new experiences and soon met someone whom she would not have noticed before. He was not what she would have imagined as an ideal partner for herself. Yet in time, as their love kept growing deeper and the relationship more fulfilling, it became clear that they were very well suited to one another. The last I heard was that they were married and considered their partnership an ongoing adventure.

Patricia took some time to shop the dealerships before she found the brand that suited her. Good for Patricia! We recommend you do the same. It's quite easy to be persuaded by others to buy a model that really isn't us, in cars and guys. Kimberly once bought a car, a very nice car, which her husband Jeff thought was perfect for her. Knowing she was in the mode for a new car, Jeff wanted to help in the search. When he came home one evening, he began telling her about the fine car he'd driven. A Mercedes E320. He just knew it was right for her, and it was a "good deal." Caught in his excitement, she was ready to see the car. She drove it for a test drive on the freeway and agreed it was a great car. So Kimberly bought it. She bought it at the same time she had decided to go back to school. It didn't take her long to realize this was not the car for her at this time in her life. She saw herself as (and was) a student, and this car wasn't a student's car. She felt a dissonance. The Mercedes was classy and fun to drive, but it just wasn't the kind of car she wanted to drive to night classes across town and fill up with book bags and junk that go with college life. It wasn't the right car for her. Six months later, Kimberly traded in her gorgeous car for a Jeep, just as years before, she had traded in a bad husband for one that was made of good husband material. You have to know what you need and be willing to stick to your guns regardless of what others think—in cars and husbands.

Whatever you do, don't settle. Our friend Maria lamented on her thirty-fifth birthday, "I'll never find the right man, and I'll be alone the rest of my life." Maria is not unlike Lucinda. She has a lot going for her—good looks, a talent, her own successful business, a sense of humor, friends, an adorable dog named Honey. Just no husband. A girlfriend asked Maria to go out to dinner with her on her birthday. Afterward the friend wanted to go dancing and troll for guys. Maria couldn't think of anything she would rather do less. But she agreed. They met two guys there. Sonny had been dragged there by his friend Ben. Ben and Maria's friend danced. Maria and Sonny talked about how much they didn't want to be there. Then they started talking about each other. They have been talking (among other things) ever since. "He's everything I ever wanted in a man," Maria says. "Honey loves him. When I get up off the sofa, she jumps up in my place next to him. He even has angels in his apartment. Can you believe it? A guy with angels hanging in his apartment! You can't settle. You have to wait for the right one." Don't try to resolve your dissonance about a man by telling yourself any kind of stories.

Here are some of the stories (as in lies) you might be tempted to tell yourself:

Lie # 1: He'll change.

Lie #2: I can live with it.

Lie #3: It'll be OK because we love each other.

Lie #4: That's how men are.

Lie #5: I'm too picky.

Don't do it! Don't believe any of these stories. And don't con yourself into accepting a relationship that doesn't work for you by buying one (or more) of these stories. Psychologist Jean Houston tells this about a friend of hers who has an elaborate set of lies she tells herself.

> I have a friend who is getting herself into a lot of trouble in a relationship. She has so many brilliant justifications. She can quote Freud, Nietzche, Kierkegäard. She has built up a tower of illusions. We all feel as if we're sitting at a play of *King Lear* just as King

Lear is about to banish Cordelia. We feel like saying, "Don't do it. She's the good one. It will wreck your life. Don't run off to Regan and Goneril." But you can't stop the play. That's the way I feel about my poor friend. We all know she's headed for utter disaster, but there is nothing we can do about it because her intellect is so extraordinary that she has built up these mammoth ways of justifying what is exceptionally stupid behavior.

("Front Porch: A Conversation with Jean Houston," by Scott London, *The Salt Journal,* Vol. 1, #1, Nov./Dec. 1997, p. 56)

If you are ambivalent about being married, live with the ambivalence until you can figure out what the problem is. The problem may be your history, a problem in the brand of husband you are test driving, or a mismatch between you and the brand. It could well be that your ambivalence is right on. Marriage to this guy or any guy at this time in your life may be wrong for you. Marriage is a good thing only if it's right for you. Remember that there are many modes of transportation down the highway of life. There's public transportation, leasing, borrowing, biking, and even walking. Ownership is only one option. You don't have to have a husband to get where you are going in life. But if you want one, read on... In the next chapter, we'll tell you how to choose a husband.

## Four

## HOW TO CHOOSE THE RIGHT BRAND HUSBAND FOR YOU

HERE'S A STORY that appeared in *The New York Times* about how one woman went about choosing the right brand man as her husband.

> Tse-Feng Susan Chang broke up with her boyfriend and concluded she had about as much chance of finding long-lasting love in New York as she had of spotting a starfish in the Hudson River. To cheer herself up, she took saxophone lessons, planted aromatic herbs in the window box of her Manhattan apartment and threw herself into ballroom dancing. She also placed a long, poetic ad in the personals section of Yahoo, an Internet service.
>
> The ad began: "I'm a young woman with an old soul; I like the hot sun and the city

lights, the glamour of the ballroom and the silence of the bookshelf. I know the difference between anise and dill, among other things, and I *almost* know what fenugreek is. I believe in Manhattan without therapy, the contagiousness of joy and the basic sweetness of life. I am looking for someone who's made the most of what came in the box: a man with a big heart and a big and open mind."

Within moments she began receiving e-mail responses, and over the next few weeks she heard from 107 people. "There were some complete sex maniacs," she said. "There were some nice, normal people. There was more than one businessman who was in town for the weekend and logged in from his hotel room. None were quite right. It was like Goldilocks—too hot, too cold."

Buried among the responses was one response that seemed so right it gave her goosebumps. It was Randall te Ve. Like her, he plays the sax, places some faith in tarot cards, enjoys ballroom dancing and has a sly, intellectual writing style. His reply read, in part: "I'd enjoy starting a correspondence with you, although my qualifications don't come close to meeting your demands. I haven't met very many who have even sorted out their box, let alone having made the most of its contents. Mine is in astonishing disarray, but I keep hope alive."

They began an on-line relationship that quickly became intense. "By the third e-mail, I was in love," Ms. Chang said. She added: "It was so great to have someone to share all your thoughts with, someone you didn't even know. We likened it to having someone to confide in behind a dark screen."

They rarely discussed meeting in person, though. So Ms. Chang devised a plan. She knew he took a course on Tibetan Buddhism and one evening slipped into the class to search for him. She knew his height, his weight and his eye color—hazel. As everyone was leaving, she spotted someone she thought might be her pen pal. "I just put my two feet down in front of him and asked, 'What's your name?'" she recalled. "His eyebrows shot up over his glasses, and he said, 'Randy.' And I said, 'Oh, I'm Susie.'"

One year later they were married aboard a wood-paneled barge with the ambiance of a log cabin, moored under the Brooklyn Bridge. "Maybe the Internet romance is a commonplace story these days—in a way I hope it is," Ms. Chang said in an interview with *The New York Times*. "Like so many women in New York, I'd given up my expectations of finding anyone. Posting a personal was really just a way to tempt fate."

(Lois Smith Brady, "Susan Chang and Randall te Ve," *The New York Times,* June 28, 1998, 7)

Susan Chang did more than tempt fate. She knew herself well enough to figure out what she wanted in husband material, and then she advertised for that brand. You know that only *you* can figure out what you need in a husband. It's like figuring out what you want to eat. A carrot stick may be exactly right sometimes, but there are other moments when only a Dunkin Donut will do. What you need now in a husband may not be the same as what you sought (or got) at an earlier period in your life. Fortunately, individual taste in husbands isn't as mercurial as taste in food, but it is entirely unique to you and completely up to you to figure out. Only you can know who meets your compatibility quotient. Compatible means "capable of existing together in harmony." Some of the brands we'll discuss would make that tough for any woman, but you know what you need to give you a sense of harmony in a relationship with a man. Helpful waitpersons, friends, and family will have recommendations about the "daily specials." The selection is up to you because you have to live with what you order—for life, unless your choice makes you too sick to stay at the marriage table to the end.

## FIGURING OUT WHAT YOU ARE LOOKING FOR IN A HUSBAND

Knowing what you *want* in a husband is different from knowing what you *expect*. As we explained in the last chapter, our expectations are a product of our experience. The kind of husband(s) your mother had is part of that experience. So is your relationship with the man who served in the father role for you. Growing up without a father plays into your expectations just as much as having a close relationship with a dad. We talked about expectations based on fathers in Chapter 3. Your mother and your relationship with her plays an equally important role in forming your expectations and wants in a husband as does your father. This is why.

We all seek what is familiar. Our experiences with both our mothers and our fathers combine into unconscious expectations of what we need in a husband and what marriage is going to be like for us. When you go shopping for a car, you have in mind a general picture of what you realistically expect to get. You figure a Toyota Camry is what you ought to get. Just for fun you may spend some time looking at Beemers or Mazda Miatas, models or brands that are way out of your price range or too impractical. But your serious shopping and negotiations will be for the model that most closely matches your image of the "right car" for you.

If your experience growing up was that cars break down a lot and are a big hassle, your *unconscious* expectation is that the car you end up with will break down and give you lots of problems. What you *want* is a car that is reliable and predictable and appropriate to your needs. What you go shopping for in husbands is a composite picture resulting from your relationship (or lack of) with your parents and their relationship(s) with each other. It also is a result of what you need from having been brought up by those people. If your father was an alcoholic and your parents fought a lot, what you *expect* to get in a husband is likely to be someone who is unpredictable and does things that upset you. What you *want* is a man who is reliable, predictable, and considerate of your needs. Your mother might have been that person in your life—

the strong, predictable one. You may *want* a man who is like your mother but *expect* one who is like your father. Our wants and our expectations of what we'll get in a husband may be opposites. If you don't know what you expect, it's likely that you will get what you *expect,* not what you want.

One of our goals in writing *The Consumer's Guide to Husband Material* is to maximize the odds of getting what you want when you choose a husband. To choose the right brand husband for yourself, you need to know the differences (and similarities) between what you expect to get and what you want to have. Once you are more aware of what makes up the model of the "right man" for you, based on your experience growing up, you can better understand why you are attracted to certain kinds of men and you can evaluate whether or not that brand of man fits your needs now.

We don't want you to be like Cinderella the day after the ball, waiting, broom in hand, for the man you remember dancing with the night before to show up, with the slipper. If your dad was fantastic and your relationship with him was loving, great! That's what you expect in a husband, and you can choose that in a husband for yourself. The emphasis is on the word *choose*. It's a dangerous thing to lie around like an unconcious Sleeping Beauty waiting for the "right" guy's kiss to wake you up. It works a whole lot better for you to be Belle in *Beauty and the Beast*. Go into it with your eyes open. You may not like what you see when you look at your unconscious expectations of what you'll get in a husband. That's OK. Be a bit horrified at first. Then you can, with the help of our rating chart, decide for yourself what's handsome and right for you and which brand of men are still going to be frogs no matter how many times you kiss them.

To help you through this crucial process of analysis, we have developed a questionnaire that will give you a clear picture of what you *expect* and what you *want*. It has two parts. Please take the time to fill out both parts. DON'T SKIP THIS STEP! It is essential in order to truly reap the benefits of this book. The questions may bring up difficult emotions, but muster your courage and curiosity and fill in all the answers. The first section of the questionnaire will help you get a picture of your image of what husbands and marriage are going to be like for you. The second section is your job description for a husband. Please do both parts. Writing your answers in a notebook or on separate sheets of paper, please answer all the questions. Here's the first questionnaire.

## WHAT AM I LOOKING FOR IN A HUSBAND?

### YOUR RELATIONSHIP WITH YOUR FATHER

1. Describe your father as he was when you were growing up (before age 12): If your father died before you were twelve, write down the description of what you remember or how others in your family described your father. If a stepfather or some other man was the main male caregiver in your early life, describe him. Use that person in your mind as you answer each of the questions that relate to fathers.

2. Describe your father's relationship to you.

3. Describe the nicest memories you have of your relationship with your father.

4. Describe your most painful memories of your relationship with your father.

5. Describe the pleasant feelings you have when you remember your relationship with your father.

6. Describe the unpleasant feelings you have when you remember your relationship with your father.

7. Describe what you most wanted from your father and didn't get.

### YOUR RELATIONSHIP WITH YOUR MOTHER

1. Describe your mother as she was when you were growing up (before age 12): If your mother died before you were twelve, write down the description of what you remember or how others in your family described your mother. If a stepmother or some other woman was the main female caregiver in your early life, describe her. Use that person in your mind as you answer each of the questions that relate to mothers.

2. Describe your mother's relationship to you.

3. Describe the nicest memories you have of your relationship with your mother.

4. Describe your most painful memories of your relationship with your mother.

5. Describe the pleasant feelings you have when you remember your relationship with your mother.

6. Describe the unpleasant feelings you have when you remember your relationship with your mother.

7. Describe what you most wanted from your mother and didn't get.

8. Describe the relationship between your father and mother.

## YOUR HUSBAND PROFILE

Now you put all this information into a composite picture of what you are expecting to find in a husband. Here's how to do it.

1. Make a list of all the adjectives you used to describe your father in Question 1 and your mother in Question 1. Say, for example, that

your father was friendly, hardworking, unpredictable, and a clown. Your mother was quiet, loyal, hardworking, and serious. Your list would be: friendly, hardworking, unpredictable, clownish, quiet, serious, and loyal. Don't worry about contradictions.

2. Make a list of your father's and your mother's relationship with you. For example, if your father was absent a lot but playful and attentive with you when he was around, and your mother was critical and never satisfied with what you did, your list would be something like this: absent, playful when around, critical, never satisfied.

3. Make a list of your nicest memories with your father and mother. It might be something like this: Dad helping me learn to ride my bike without training wheels and taking me with him to his office on Sunday afternoons; Mom holding me in her lap and reading me a story.

4. Make a list of your most painful memories of your relationship with your father and your

mother. An example would be: My father not being at any of my birthday parties and not being around for many of the important events in my life; my mother never praising me directly and always comparing me with my older sisters.

5. Make a list of the pleasant feelings you have when you think about your father and mother. Possibilities could be something like: happy when I remember how funny my dad was and excited when he was going to come home; proud when I think how hard my mother worked to set high standards for me.

6. Make a list of the unpleasant feelings you have when you remember your relationship with your father and mother. The list might be: lonely when I remember how much I missed him and angry when I think about how I could never please her.

7. Make a list of what you most wanted from your father and mother that you never got. Examples could be: his attention, her approval.

8. Make a list of the adjectives that describe the relationship between your father and mother. The list might be: distant, cooperative, cordial.

You've worked hard. Now just put it together in writing (or even reading it out loud to yourself if you hate to write.) This isn't difficult. Stick with it a little bit more. Here's what to do. Just fill in these sentences with what you have already written.

1. The man I expect to marry is (fill in with your answer to Question #1 on "Your Husband Profile").

2. The man I expect to marry will have this kind of relationship with me (fill in with your answer to Question #2 on "Your Husband Profile").

3. The kinds of things my husband and I will do are (fill in with the types of activities you listed in answer to Question #3 on "Your Husband Profile," such as playing cards, going for walks, etc.).

4. What would hurt me the most is if my husband (fill in with your answer to Question #4 on "Your Husband Profile").

5. What would make me most happy is if my husband would (fill in with your answer to Question #5 on "Your Husband Profile").

6. The bad feelings I expect to have with my husband are (fill in with your answer to Question #6 on "Your Husband Profile").

7. What I most hope to get from my husband is (fill in with your answer to Question #7 on "Your Husband Profile").

8. What I expect my relationship with my husband will be is (fill in with your answer to Question #8 on "Your Husband Profile").

## WHAT-I-HAVE-TO-HAVE-IN-A-HUSBAND JOB DESCRIPTION

You have just completed the hard work of figuring out what brand of husband you are looking for—whether you knew it or not. He's the man who will be most likely to fulfill the expectations you have so carefully described. You might not be so delighted with what you see. If you had a father who cheated on your mother, that, as you have just seen, is part of your expectation of the man you will choose for your own husband. If your expectations, based on your history, paint a much rosier picture, in all likelihood, you are shopping for—and will find—a man who can help fulfill that happy expectation. Regardless of what your expectations are, you are now much more informed about who you are shopping for, why, and what kind of classified ads you are responding to when you go checking out what's available.

Now comes the fun part. You get to decide what your *ideal* husband would be. Write a description of what you ideally would like to have in a husband, what you really would want, not some imagined image. A movie star might be perfect for some women but if the idea of living with one terrifies you, don't describe a movie idol as your ideal husband. Be serious about this. It really is an order for your dream husband that you are placing. It's the "job description" for your husband. List everything that is an essential feature—what you *have* to have in a husband. After that, list optionals—what would be nice, but you could live happily without that feature. Fill it out as completely as you can. You can go back and add or modify later.

As you can see by now, you have been doing comparison shopping whether you knew it or not. You have been comparing your husband candidate with two images: the image of your experience with the original most-important-in-your-life people and with an idealized image of who you think would make you happy ever after. The two images may closely match each other, or they may be total opposites. If what you got psychologically from your parents was great, your WHAT-I-HAVE-TO-

HAVE-IN-A-HUSBAND job description is probably a pretty close match to your description of your parents and your relationship with them. But if growing up was rough, your job description of WHAT-I-HAVE-TO-HAVE-IN-A-HUSBAND is likely to be what you hope will be a healing relationship. When you look at the brands of husbands we will discuss in the next chapter, look carefully to see which brand man most closely fits your expectations and/or job description. If your father or mother was an alcoholic, read about The Addict because that's the brand you will be expecting to take home. Also read about The Eagle Scout because he may more closely fit your job description of an ideal husband. The point of all this, as we said at the beginning of this book, is to get you to *think* about the choice you are making in deciding what brand of guy you will make your own. We want you to marry *for better,* not for worse. The work you have done here will help make that happen.

In Chapter 5, we describe different brands of men and the basic features of each. Read about all of them. You might decide to look into a brand you hadn't considered before. Keep an open mind, even if you have already signed a lease/purchase agreement with someone. Remember, it's your choice which lucky guy gets to be parked in your garage.

## Five

## BRANDS OF HUSBAND MATERIAL

IN ORDER TO MAKE your research as simple as possible, we have categorized men into sixteen brands. The brands are: Addict, Big Daddy, Bubba, Bully, Couch Potato, Eagle Scout, Jock, Macho Man, Mama's Boy, Narcissist, Outdoorsman, Peter Pan, Poet, Power Seeker, Romantic, and Workaholic.

Few men will be only one brand. Just as some Toyotas have engines as powerful as the Corvette, an Eagle Scout may also be a Power Seeker. Some men may have characteristics of many of the brands. That's not necessarily bad. It can mean he's a well-rounded person not bound by any stereotype, or it can mean he's a big mess. You'll have to come to that conclusion yourself. By reading the descriptions of each brand you will be able to identify your guy. Then look at the Rating Chart in Chapter 6 to see how each brand man ranks as a husband on the twelve essential features. A word of caution: Just because a guy drives a certain kind of car doesn't mean his basic brand is like the car he owns. You have to look at the description of the guy, not what he tootles around town in. Here are the brands. Start your research and find the description(s) of your potential husband material.

## THE ADDICT

The addict is a guy with an accelerator that sticks and brakes that fail. You never know when either will happen, but when they do, the ride can be really scary and possibly fatal. Crashes are an ever-present possibility. Where the accelerator sticks varies. For some men, it sticks on sex. They become sex addicts. Others stick on alcohol or drugs or food. One addiction, work, is its own brand, because guys with this addiction are a different model in some ways than other types of addicts.

If your husband-brand-of-choice is an addict, you had better have a pretty high tolerance for unpredictability and a loose grip on the steering wheel of his life. He may be loving, sweet, considerate, and a hardworker, but you never know when that accelerator is going to jam and you'll find yourself on Mr. Toad's Wild Ride.

Addicts can be good husbands when they aren't in the grip of their addictions. We know one wife who accommodated to her husband's alcoholism by sending him to a hotel if he came home drunk during the week. "We've got homework to do here, and you're getting in the way," she bluntly told him. He had done enough recovery work, from which he often lapsed, to know that was how it had to be. So he left until he sobered up the next day. Life with addicts, as with any husband, is a compromise. The compromises you have to make to have an addict brand as your man may work for you and they may not.

Addictions are about pain. When the addict brand husband hits a pothole of pain in his heart and mind, the accelerator slams down into sex, drugs, booze, or food in order to lurch him past the pain as fast as possible. Those addicts who have done extensive repair work on their accelerating and braking mechanisms learn to maneuver past painful potholes and to take detours onto less rough roads. That's *his* work to do. Not yours. If he does the work, the ride will be more pleasurable for both of you. If he doesn't, you may have to do your own hard work to keep your hands off his steering wheel. Grabbing the steering wheel of someone else's life is a very dangerous thing to do.

## THE BIG DADDY

The Big Daddy needs a little girl. A good little girl. Good as in being good to Daddy. His image of himself is a Lincoln Town Car: a big, old-fashioned rear-wheel-drive highway cruiser, with a powerful engine and lots of luxury appointments. He wants you riding in that big front seat beside him. He doesn't expect nor want you to take a turn behind the steering wheel of your relationship. We know a Big Daddy who presented the income tax return to his little darling with only the signature line exposed for her to sign. No need to trouble her pretty little head with boring old numbers.

Having a Big Daddy model in your garage can be a luxurious ride, but the emphasis is on "ride." You have to be willing to ride along with the game that he's the parent and you're the cute kid. He gets the new Town Car, and you get to drive his old one. This means you can't act like a grown-up woman who knows she can manage fine without him, which may be no problem if that's how you see yourself.

Big Daddy likes to buy his little woman playthings—toys that will keep her laughing and smiling. Diamond tennis bracelets, sporty little expensive cars, fluffy fur coats, and whatever else he thinks will make his little doll happy. All these playthings come with a price, of course, and the price is an agreement to let him be in charge of your life. If that sounds great to you, go for it. Besides, you may be able to rationalize the arrangement by convincing yourself that it's actually you who are running his life. There's nothing inherently bad about the Big Daddy model, but when you are living life limited to a role—any role—trying to be authentic can be cumbersome.

Trying to fit into your Big Daddy's fantasy can get sinister if the fantasy includes a requirement that your body meet some Barbie Doll specifications. If you think you have to change your shape surgically to be what he wants, consider a trade-in immediately. You can have permanent lip color or permanent eyeliner tattooed on you. You can have a boob-job, a tummy tuck, a face-lift. It won't matter. You won't be perfect enough because you will still be a real person with real flaws. If your Big Daddy can't love you flaws and all, shop for a man who will.

## THE BUBBA

The Bubba is likely to be found, in fact or in his dreams, driving a pickup with a gunrack in the back. Not much subtlety here. He's simple (but not necessarily simple-minded) and straightforward. Like a pickup, he's a rugged and obvious sort of fella who can give a woman a sense of being protected. With one of these models, you may feel comforted in knowing that he would be quick to defend your honor and protect you from potential predators. For some Bubbas, his mission is protecting his woman, which is great if you feel a need for that sort of security. The drawback is that his attitude of protectiveness may start feeling like possessiveness, which can obviously be stifling if left unchecked (see Bully below).

Wearing boots and jeans and a big belt buckle (the chaw in the mouth is optional), this model man is basically a big kid. Another version of the Bubba wears camouflage clothes and drives an old army truck up mountain trails marked "No Motorized Vehicles." Don't let the costume confuse you. Custom detailing doesn't change a basic model. The Bubba is fun, in a crude sort of way, no doubt about it. If you like watching monster truck races, you'll have fun with a Bubba brand of guy. Kids *are* fun. The problem comes when it's time to stop playing. He needs his woman like a ten-year-old boy needs his momma. He'll come to you for food, sex, fun, fighting—but not much of just hanging out or helping out.

The Bubba model can communicate fighting words and basic needs easily, but subtleties like "I love you" and "I appreciate you" come hard for him. Intimacy is not a standard feature on this brand. No matter how this Big Ram is dressed up, he never will let you forget he's a truck. He has his dog sitting next to him in the truck, and you'll likely be in the back. If you're willing to be the Big Momma to this Bubba, he may be just the model for you.

## THE BULLY

Think of the Hummer car when you think about the Bully brand of men. The Hummer is the vehicle that gained fame in the Gulf War for its toughness in all sorts of conditions. It takes up more than its share of space and gas, and definitely makes a statement: "Get out of the way! I'm in control. I'm the boss man." Remember, this vehicle was designed for war, and so is the Bully. He's a psychological *Exterminator*. He may not look like a battle-ready soldier, but he's quick to start fights and use aggression, mental or physical, to get his way.

Margaret hated riding in the car with her husband Doug. Doug was going to mow over anyone in his path. He would drive up on the esplanade to pass cars waiting for a light to change. He'd speed through neighborhoods and disregard traffic signs with impunity. Doug drove the way he lived—as a bully. Not all bullies are as obvious as Doug, but this brand of man as a husband will give you a bad ride with awkward handling.

Like the Addict, the Bully has poor control of his anger brakes, and, to make matters worse, likes to use his engine power to scare others. If you pick a Bully for your husband, you will be signing on as a badly treated passenger or chauffeur in the vehicle of his life. Where you want to go and how you want the ride to be in your life will become irrelevant to him.

At this point, you might be asking in disbelief: why would anyone want to buy a Hummer guy? The maintenance is very costly, and the MPG is lousy. But if you grew up with bullies, as you already know from the work you did in Chapter 4, you can find yourself shopping for a brand of guy whose handling is comfortably familiar.

Also, like the Hummer vehicle, the Bully has some redeeming qualities. He doesn't score all F's. No guy does. He may be very committed to you, in that he is counting on you to make his life run well. If taking care of a man's every need is what you have as your plan for your relationship with your husband, perhaps you shouldn't eliminate the Bully brand from your list of possible candidates. However, as you will see in the rating scale, a Bully is certainly not a contender for any Husband-of-the-Year award. He may take you on a vacation, but it's not going to be any vacation for you. Plan on taking care of him and any kids you might have with him, but don't plan on his taking a turn at taking care of you. If you do get him to carry the cargo for a while, it's going to take so much work on your part to get everything ready that you are likely to end up saying, "Forget it. I'd rather do it myself."

The Bully is likely to have grown up in a family that was a school for aggression. As Daniel Goleman, author of *Emotional Intelligence* (Bantam Books, 1997, pp. 196–197), explains:

> As children, the troublemakers had parents who disciplined them with arbitrary, relentless severity; as parents they repeated the pattern. This was true whether it had been

the father or the mother who had been identified in childhood as highly aggressive. Aggressive little girls grew up to be just as arbitrary and harshly punitive when they became mothers as the aggressive boys were as fathers. And while they punished their children with special severity, they otherwise took little interest in their children's lives, in effect ignoring them much of the time. At the same time, the parents offered these children a vivid—and violent—example of aggressiveness, a model the children took with them to school and the playground, and followed throughout life. The parents were not necessarily mean-spirited, nor did they fail to wish the best for their children; rather, they seemed to be simply repeating the style of parenting that had been modeled for them by their own parents.

In this model for violence, these children were disciplined capriciously; if their parents were in a good mood, they could get away with mayhem at home. Thus punishment came not so much because of what the child had done, but by virtue of how the parent felt. This is a recipe for feelings of worthlessness and helplessness, and for the sense that threats are everywhere and may strike at any time.

So you can see that the Bully brand of man has structural design problems that will make the ride rough at best, and dangerous at worst. If his bullying becomes physical, don't think any further of taking this man as husband material. (Review disqualification's in Chapter 2.) Staying with a man who is physically aggressive to you is like continuing to drive a car with smoke billowing out of the hood. Get out of his life as fast as you would jump from a burning car. You can't stop to think about the loss. Just run for safety.

## THE COUCH POTATO

This is the Archie Bunker model of man, a jalopy kind of guy. We've all watched the sit-coms that show a balding-on-the-top, bulging-in-the-middle guy eating, drinking, and scratching in front of the television set. We've heard stories from our friends about their husbands, who park themselves in front of the TV and only have sufficient combustion to do the refrigerator/bathroom shuffle. This model gets poor mileage as a result of consuming low grade octane, but he takes up quite a bit of space because he is often in the reclining position.

The Couch Potato may look similar to other models at first glance. In this age that esteems speed and activity, these men have learned how to camouflage their propensity for slipping into idle. We, however, have put together some signs to help you identify a true Couch Potato.

- Check out where he parks himself. Does the big Lazyboy near the big screen TV look well-worn? A sure warning sign.
- How many sports channels does he subscribe to? More than one means he really likes to have many options to keep himself well-entertained during the many hours he plans on being stationed in front of the TV.

- Can he change channels without looking at the numbers on the remote control?
- Does his Lazyboy come equipped with accessories like a miniature refrigerator on one side and a TV tray on the other? If so, this is the luxury model of the Couch Potato.
- Commercials are your best shot at sex, and you're out of luck entirely during the Super Bowl.

Some Couch Potatoes may be more than lazy. They may be depressed. Depression is the psychological version of a stalled engine. With the right combination of spark, fuel, and oxygen, the engine can fire up and down the road to fun you and your hubby can go. If nothing in life seems to get your Couch Potato guy sparked, see about having him towed in to a physician or therapist for a check-up. Even if it's not in his make-up to have much get-up-and-go, no one is supposed to have an engine that never turns over.

On the positive side, if you want a brand guy that's going to be parked in your carport, the Couch Potato guy is a model you might want to consider. He's not likely to jump into overdrive unexpectedly, but you might find yourself dissatisfied with the effort it takes to get him in gear and rolling down the road of life with you.

## THE EAGLE SCOUT

This brand man has an unstinting array of standard features. He's a Jeep Grand Cherokee Laredo. To understand the Eagle Scout model of man, it's good to start by knowing what scouts promise to do in life. This is the Scout Oath:

> "On my honor I will do my best
> To do my duty to God and my country
> and to obey the Scout Law;

> To help other people at all times;
> To keep myself physically strong,
> mentally awake, and morally straight."

So what is "the Scout Law?" Actually, it's pretty good instruction for the making of good husband material. Here's the Scout Law:

A Scout is TRUSTWORTHY. A Scout tells the truth. He keeps his promises. Honesty is a part of his code of conduct. People can always depend on him.

A Scout is LOYAL. A Scout is true to his family, friends, Scout leaders, school, nation, and world community.

A Scout is HELPFUL. A Scout is concerned about other people. He willingly volunteers to help others without expecting payment or reward.

A Scout is FRIENDLY. A Scout is a friend to all. He is a brother to other Scouts. He seeks to understand others. He respects those with ideas and customs that are different from his own.

A Scout is COURTEOUS. A Scout is polite to everyone regardless of age or position. He knows that good manners make it easier for people to get along together.

A Scout is KIND. A Scout understands there is strength in being gentle. He treats others as he wants to be treated. He does not harm or kill anything without reason.

A Scout is OBEDIENT. A Scout follows the rules of his family, school, and troop. He obeys the laws of his community and country. If he thinks these rules and laws are unfair, he tries to have them changed in an orderly manner rather than disobey them.

A Scout is CHEERFUL. A Scout looks for the bright side of life. He cheerfully does tasks that come his way. He tries to make others happy.

A Scout is THRIFTY. A Scout works to pay his way and to help others. He saves for the future. He protects and conserves natural resources. He carefully uses time and property.

A Scout is BRAVE. A Scout can face danger even if he is afraid. He has the courage to stand for what he thinks is right even if others laugh at him or threaten him.

A Scout is CLEAN. A Scout keeps his body and mind fit and clean. He goes around with those who believe in living by these same ideals. He helps keep his home and community clean.

A Scout is REVERENT. A Scout is reverent toward God. He is faithful in his religious duties. He respects the beliefs of others.

Sounds perfect, huh? Well, these are *laws,* and laws are standards to shoot for. It's helpful, though, to have a guy who aims high. Whether or not he hits the bull's-eye every time, he's going to be a more proficient marksman in the art of husbandry if he has as his goal achieving mastery in these areas. But remember that no one is perfect, and even if the Eagle Scout guy were perfect, he may not be perfect for you. If you like walking on the wild side, choosing an Eagle Scout brand could be a problem. He might serve to balance your freewheeling ways, but if you want to zip down the road of life in a Mazda Miata don't shop for a Dodge Caravan. It's not a man's job to balance what's missing in your make-up. If you need controls, don't look for them in husband material. Develop them in yourself. On the other hand, if you need/want a man who basically lives by the Scout law, this guy is a good one to consider.

The Eagle Scout tries to do everything well, and he may focus on the external to the detriment of internal awareness. "Shoulds" count a lot, sometimes overriding wants and needs when wants/needs are what's most important at the time. For example, he won't want to rock the boat. If you are having emotional problems, he is likely to try to do what he thinks will make you happy, but he may not easily be willing to dive down into the pit of painful feelings. Ultimately, chaos happens in every life. There are times when hard choices have to be made that seem to go against the Scout Law.

Remember the old joke about the boy scout who was trying to help the old lady cross the busy intersection? Trying to "do a good turn daily" as the Scout slogan advises, he offered his assistance. The old lady declined, but the boy persisted and dragged the old lady to the other side. "There!" he said proudly. "I told you I would get you across safely!"

"Yes, you did," she answered, "but I didn't want to be on this side of the street."

By being so intent on doing a good deed, the scout missed the point of what would really be helpful. So just make sure your Eagle Scout knows where you want to go in life, and you'll both be happier.

Finally, if you find a true Eagle Scout, appreciate the fact that he earned that status. To achieve Eagle Scout status requires lots of work and mountains of merit badges. So he has indeed learned much that qualifies him as good husband material. The caveat above is just so you know that merit badges don't make a perfect husband.

## THE JOCK

The Jock is the muscle car of men. The Chevy Camaro. As *Consumer Reports* describes, "Seating is decent for two but the driver cannot see out well." He has a narrow view of where he's going. Whatever age he decided to make his muscles his priority is the age when he stopped growing emotionally. There is a difference between the athlete and the jock. The athlete is one who sees a sport as an avenue for personal growth and challenge. Sport is as much mental as physical for the athlete. A Jock brand of guy is going for show. He has a rigid body and a limited interior. The body is busy, with awkward flexing. Honda had one like this called the del Sol, but it discontinued the car in 1997. Makes you think, doesn't it?

Performance matters to the Jock. He may work as hard at perfecting his sexual skills as he does in building his pecs. Enjoy that part of this brand, but don't expect much pillow talk afterwards. We know a woman, who when she was researching husband material, tried out a Jock model. Her requirement of him was that he not talk. "Just sex," was her dictum to him. She wisely had a lease arrangement and decided at the end not to pick up the option to buy. Again, it all depends on what you are in the market for.

One of the strengths of the Jock is his reliability. That may seem contradictory with his wandering potential, but the Jock is not a complicated brand, and he's capable of showing up for scheduled practices and workouts. Like the Couch Potato, you can pretty much know where he'll be (at the fitness club) and what you can count on him for doing. On the other hand, he's only up to handling his workout sessions and isn't likely to want to set aside his routine to help you with broken equipment.

This brand needs lots of reassurance that he's terrific, but it may not occur to him that you need that as well. Susan loved her Jock Len for many years, and he adored her—and how great she looked. They met at the fitness center, of course. Len was seven years younger than Susan, which was OK with her except when they would go out with his friends and their even younger girlfriends. "Look cute tonight," Len would cheerfully instruct Susan as they made plans for meeting the friends.

Looking cute for Susan, who had pushed past forty a few years earlier, was doable but it got to be more of an irritant than a challenge she was interested in meeting. She's hopeful that her cuteness becomes less important to Len as they both age and that he'll love her no matter what. We hope so, too, for her sake and his.

One other problem with the Jock is that if he's not getting what he thinks is sufficient praise from you, he is at some risk of not waiting too long before looking for it elsewhere. On the other hand, if he's basking in your adoration, he may adore you. Think of a Ken doll, and you got the idea of what's involved in maintaining a Jock. You have to direct the Ken doll, lead him, talk for him, and after all that, the bottom line is he's still plastic, stiff, and rigid.

## THE MACHO MAN

A group of men calling themselves the Macho Movement named President Clinton "Macho of the Year" for 1997. "There is no question that, considering Clinton's performance in 1997, no one did more to deserve the award," group president Luis Mario Ladeira gushed. "No other public figure in 1997 did such honor to macho traditions. With all the revelations, we became aware he is a true macho." Previous honorees include former Brazilian President Itamar Franco, selected after he was photographed dancing with a model dressed only in a wet T-shirt at the Rio de Janeiro carnival. (*Houston Chronicle,* Feb. 3, 1998, 2A) We all know what else made Clinton so deserving of that award.

The Macho Man's strongest selling feature, at least in his mind, is his dick. He's interested in it, so he thinks you are, too. And, perhaps you are, but the part definitely doesn't make up the whole. Eventually, the novelty of that feature does wear out, and you probably will discover sooner rather than later that there are many other features you will require in husband material. This guy is a Corvette. It's legendary, and that's what the Macho Man thinks he is, too. Like the Jock, he's muscular in ego but more refined than the Bubba. This guy has sex appeal of a certain kind, and he may be exactly what you need if it's the kind that turns you on.

Macho Men are not lacking in other virtues. They may make great salesmen—or politicians—because they can be very skillful at selling their wares. Obviously, given President Clinton's dubious award, they can go far. The problem comes when they feel a need to keep selling their essential feature after you have already signed the title papers. It's a real drag to keep having to check his emergency brake to make sure he isn't rolling down the driveway and out into the street. You don't want to have to worry about which gear his stick shift is in.

It's OK if a man likes having his identity tied to his masculinity, but that's limiting if it's all he ties it to. Sexual potency is important. None of us would be here without it. Some guy had to be macho enough to help our mothers make us. There's a time and place for everything, and the Macho Man may have trouble distinguishing those times when "machismo" is inappropriate or even harmful. We all like to play a role in life. A guy who enjoys wearing the Macho Man hat is not to be faulted unless he forgets or never learned that it's just a role.

## THE MAMA'S BOY

Sitting next to us at a dinner party, Kathleen explained why she ended a seven-year relationship with a man she really loved. "I realized the relationship wasn't going anywhere."

"Don't tell me," Rita said. "Let me guess. I bet he was a Mama's Boy."

Kathleen looked astonished. "You're right! He actually went home to live with his mother and he's thirty-nine years old and has a job as a geophysicist!"

A Mama's Boy can make great husband material if you can ever get him away from his mama. He's loyal, wants to please, and is basically a good person. But if the mama still has her apron strings wrapped around his neck, this brand guy will always be a loaner, and you'll never get title papers. Some mamas can keep their boys tied up even from the grave.

That's why we rate this brand guy F/A under Commitment (see Chapter 6). Until he has broken the emotional symbiosis with his mother, he can't commit to you. If he does, he's yours to the final breath. The same dichotomy applies to maturity.

The Mama's Boy gets a D/A in Maturity. It takes maturity to be psychologically independent of our mothers. Once he's done that homework, he goes to the head of the class in this feature.

A Mama's Boy sees his mama in every woman he loves. That's bad and that's good. Forrest Gump was a lovable Mama's Boy. He always did what she taught him was right. She gave him simple philosophies to live by, such as "Life is a like a box of chocolates. You never know what you're going to get." The Mama's Boy may see you with a child's innocent eyes, which can be sweet. But if he hasn't really found psychological independence from his mother, he may rebel, as an overcontrolled child would do.

Benjamin Spock, the famous baby doctor, was a Mama's Boy. His mother, Mildred, once faked a heart attack to get Benjamin to cut short a visit to his fiancée, Jane Cheney. He never broke free from his mother psychologically. Jane, whom he married despite the *coitus interruptus* his mother caused, was a dominant woman like Mildred. Spock rebelled against her in a way he never did against his mother. This rebellion took the form of flirting. He worked the room like a gigolo, ignoring Jane, who would sit in the corner, drinking herself into a mean tizzy. (*Dr. Spock: An American Life,* Thomas Maier, Harcourt Brace and Co., 1998) Mama's boys can be famous, but they can't be yours as long as their theme song is "My Heart Belongs to Mama."

So how do you know if he has broken free of this maternal bondage? You can't know necessarily by what he's accomplished in life. In fact, Mama's Boys can be high achievers because they are never sure they are doing enough to please mama. If he seems to try real hard to please his mother but gets equally impatient with her, he may be a Mama's Boy struggling to wiggle out of her apron pocket. He's very sensitive to signs that he's displeased you, but he bristles if he thinks you are trying to control him. He wants to love you and be loved in return, but he's still working out how to do that and remain an autonomous person. When he gets those two opposing needs pretty well balanced, he can be good husband material indeed. Think of a Saturn, simple, no false presentation, reliable, no-hassle-pricing. The basics of husband material are all met with this brand, given the caveat we discussed above.

## THE NARCISSIST

If you see a Jaguar XKE with personalized license plates that say "Mensa," you can be quite certain the guy behind the wheel is the Narcissist brand of husband material. This brand man is named after a youth in Greek mythology who pined away in love for his own image in a pool of water. He has excessive admiration of himself and expects everyone else to share in this exhalted opinion. He believes the advertisement for the 1998 Corvette applies to him: "The ride of a lifetime is waiting for you. Its beauty is captivating. Its performance is electrifying."

The Narcissist is a guy who has to play king of the hill with everybody. Dr. Lawrence E. Miller, a psychiatrist who describes himself as a former "King of the Hill" in his book by that name, explains the "King," or Narcissist this way:

A King dominates by skillful manipulation, seizing on others' vulnerabilities. Though not a social chameleon or interested in "fitting in" (for this implies equality), he is often able to read those around him and adjust his behavior to every occasion. He will always seek to be the center of attention and act in a way that makes him look good to others. If he senses he can't dominate a gathering of people, he won't even try. Instead, he will flee.

Without others to command and others to blame, a domineering man cannot feel superior and thus gratified. If he does have opportunity to dominate those around him, he will use either one or a combination of several ways to assert his authority. Most often his approach involves open domination of his spouse....

*(King of the Hill: For Women Who Choose a Balanced Marriage over Male Monarchy,* Bookmakers Guild, Inc., 9655 West Colfax Ave., Lakewood, Colo., 80215, 1989, p. 28.)

While it might initially give your ego a boost to be courted by someone so impressive, your poor ego will start to take a bad beating before long. You might think you are going to be queen, but the real job opening is handmaiden. Your job becomes making him happy, which is as impossible with a Narcissist as making a sieve hold water. He just has too many holes in his own self-worth to be able to hold onto any of the love and kindness you pour into him.

The allure of the Narcissist, like the Jaguar, is tempting. For a first dinner date, he might fly you to another city. He means to impress and is skilled at doing so. One woman, with a Narcissist husband, tells people who say they enjoyed meeting her husband, "Good, because you won't like him for long." The Narcissist brand needs frequent and costly repairs. He's not a suitable brand for the long distance drive of the rest of your life if you want a dependable ride. Try to imagine yourself as an old woman climbing into a Jaguar XKE. There's not much room in it (just as there isn't much room in his heart for you), and it's difficult to get in and out with a "mature" body. Having an old woman riding next to him in the sports car of his life is not going to fit his image. Some cute young babe is more what he has in mind and that's who will be sitting there sooner or later if you have managed to hang in there to your old age. Of course, all of this doesn't matter if what you want is a fun, lease car/guy relationship. The Narcissist can be a great lover. He works at it, if being a great lover is a big part of his image of himself. But what you end up with in the long run may be a dysfunctional relationship with an erection. The relationship will probably be more fun until he thinks he owns you. So if you really want to know what it's like to ride with a Narcissist brand of husband, consider a lease/borrow arrangement without an option to buy.

## THE OUTDOORSMAN

This model man makes room for heavy loads. Think Suburban, as in family, because that's how he thinks. He's nostalgic. His great-grandfather owned a gun, which he passed down to the Outdoorsman's grandfather, which he in turn passed down to his father, and now to the Outdoorsman. Heritage is important to him and, therefore, so is family and tradition. He's a hunter and a fisherman, because he sees hunting and fishing as a way of life for millions of years, and he's just continuing that heritage. Having a wife and passing on traditions to his kids are part of his design. When he's not hunting or fishing, he's likely to be spending his leisure time with you (and the kids once you have some). But you have to figure on using a loaner car during hunting season and when the fish are biting.

The Outdoorsman respects women, but he may lapse into behaving like a Macho Man or Bubba when sitting at the deer lease or campgrounds. It's not that he dislikes women. He just regresses to a more primitive mind state in that setting. He slips into a more narrow focus. Rob Becker, in his solo comedy hit "Defending the Caveman," explains how the primitive caveman had to be able to focus in order to survive as a hunter. The Outdoorsman is the direct descendant of this caveman hunter. When he's focused on his prey, there's no distracting him. This is great if he's focused on a project around the house that you want done, but if he's got his sights on the hunting trip coming up next weekend, you might as well get busy with something else you want to do without him because psychologically he's already out on the hunt.

He communicates really well on how to get to the hunting blind and how to tie a fly, but he may stall on emotions unless it's something he's lost that has sentimental or nostalgic value (e.g., his deer lease). Bill could talk to Jimmy until the wee hours of the morning about a dream deer blind and the virtues of classic shotguns. But when his wife Kelly wanted to talk with him about their relationship, Bill stammered, "You know I love you!" and couldn't imagine that there was anything else to be said. For the most part, the Outdoorsman is mature. Romance may be limited to candlelight and his throwing a steak on the grill, but he will be wanting to have you there to share in life back in the cave once this hunter returns.

## THE PETER PAN

The bumper sticker on Peter Pan's life reads "Why grow up?" He sees no point in giving up the fun and freedom of being a kid, and the Peter Pan brand is indeed fun. He loves play and adventure. He's the cruise director on the Love Boat. He's a go cart. Zippy, out in the open, no seatbelts, and just sufficient braking and steering capacity to avert major crashes. The limitations of the Peter Pan model are obvious. You wouldn't drive a go cart cross-country or even across town unless you were desperate. No passenger room, no trunk space, very short distances without refueling. Clearly we have some serious drawbacks with this kind of guy. Once you get off the go cart track and want to do something other than have fun, like go get groceries, the ride with Peter Pan is difficult, to say the least.

If your inclination is to rescue, you may be attracted to the Peter Pan brand. Most women try it at least once. It's like taking home a stray puppy. There's such a tug at your heartstrings that you forget what a hassle this is likely to be.

Christine, a lanky blonde with short-cropped hair and a passionate response to all she encounters, found her Peter Pan Joel when she was twenty-one. He had wild hair, bad teeth, and a leprechaun face. "He's never known a loving family," she pleaded defensively to her parents and friends when they reacted with astonishment at her interest in Joel. It was Christine's intent to love poor Joel into wholeness. It didn't work as it is wont not to do (see Chapter 1), but Christine was young enough to let her Peter Pan brand fly out the window to his next adventure, leaving her feeling only a little sadder and wiser. If you're going to take a test drive with a Peter Pan guy, it's good to do it early in life as Christine did. Learn what you need to learn before the messes get too hard to clean up.

You have to plan on your own transportation if you choose a pure Peter Pan. Peter Pans need Wendys. They have to have someone else take care of the responsibilities of life—yours and his. In the Disney movie, Peter invites Wendy and the other kids to join him and Tinkerbell in thrilling adventures. Of course, the adventures become frightening and dangerous, as adventures are wont to do. Whereas Peter performs heroically in saving the day, it's worth remembering that it was Peter who got them all in the messes in the first place. And it is Wendy's job, as the responsible one, to say when it's time to go home to their safe beds again. It's great to have adventure in your life and to fly with a Peter Pan brand of guy. Life shouldn't be boring. But remember what this brand can and cannot do for you and be prepared to be the one who says, "Peter, we have to go home now."

## THE POET

The Volkswagen Beetle Bug is back. We hope Poet models of men are, too. The Lovebug, as it was affectionately called, is now revived with some charming new features, the most endearing of which is a built-in bud vase on the dashboard. This touch is symbolic of the Poet brand of husband. Like the Bug, he's unique in his styling. He may have some trouble lightening up, but, on the other hand, it's good to have a man who looks at life thoughtfully.

Jim Morrison of the 1960s group, The Doors, was a singer with a lot of fame. "Come on, Baby! Light my fire!" was Jim Morrison's credo. Despite his fame as a singer, Morrison said of himself, "I'm a poet first and a singer second." He died of a drug overdose in 1971 at the age of twenty-eight. You have to be careful with Poet models of men; make sure you don't get one that takes himself too seriously or the ride in his life may be melancholy at best and tragic at worst. Some Poets are funny, like a bright orange or lemon yellow VW Bug, but they're still serious in that they will safely get you where you need to go. Others are just practical, but still make an unconventional statement in that they don't go through life thoughtlessly. They see themselves differently than most men. They think about themselves in relationship to society, values, beauty, and creativity, more than most brands of husband material.

If you think a Poet is the right model man for you, consider Volkswagen's new ad for the Beetle: "Less flower. More power." You would do well to evaluate the weight-bearing capacity of the one you have in mind. In college, Rita had a boyfriend who drove a tan Bug. He convinced her to let him come get her for a romantic weekend at his college. It was snowing hard on the way to the tryst two hours away. Rita had to sit in the back seat the whole way so that there would be enough weight to keep the car from slipping on the road. You definitely don't want to have to ride in the backseat of your Poet's life just so that you can join him in a fantasy.

## THE POWER SEEKER

Think Cadillac and Corvette when you think of this brand guy. He's got a powerful engine and is comfortable using that power on the highway of life. Like the engine, he works hard and has what it takes to go long distances in a relationship. He expects life to be a luxurious and fun ride and is willing to make it that way. The Cadillac is his work mode. The Corvette is how he plays. Work hard. Play hard. That's this model guy. The Power Seeker, like the Cadillac and Corvette, is well engineered, with effortless acceleration, and, for the most part, very reliable. He has good automatic transmission in how he shifts gears from work, to family, to play. He likes fancy toys and is happy to have you play with him, whether you drive the golf cart or swing along beside him. He's going to play, with or without you, but he'd be happy for your company.

Even luxury brands have drawbacks, however. The Power Seeker runs some risk of taking himself too seriously. The healthy Power Seeker remembers that it's all a game, after all. The man who confuses self-worth with having power is like the teenager who believes his awesome car makes him equally awesome.

The healthy Power Seeker sees the big picture, not details. He feels when the engine of a situation is not running right. The Power Seeker isn't a brand guy made to sit in the garage. This man needs to be moving. He needs stimulation and starts to feel rusty if things are too quiet. The Power Seeker is a driver in the Indy 500, not one of the pit crew. He's not likely to notice what needs doing around the house, but if you ask for help, he's likely to be a team player.

Malcolm Forbes, financier and ranker of the Super Rich, was a classic Power Seeker. He loved to gather his pals on Harley-Davidsons and off they went. He sailed and loved fine wine. He had lots of fun. To be Mrs. Forbes meant keeping up with the fun and being supportive of the work. It can be done, but having a Power Seeker man requires that you have confidence behind the wheel of your own life.

## THE ROMANTIC

They don't make this brand of man as much as they used to. That's why we women love them so much when we see them on the big screen. They are the Karmann Ghias of men. Brad Pitt is today's Karmann Ghia. Leonardo DiCaprio is another one. Clark Gable was a vintage Karmann Ghia. In case you had the misfortune to be born or grow up after the time of Karmann Ghias, we'll tell you about them. They were sleek, smooth, flowing, and made for riding along the coast with the top down and sun warming your face until the stars popped out in the dark night sky. Volkswagen, who made the car, tried to convince consumers to focus on the beauty of the Karmann Ghia and ignore its small engine.

This brand man loves a fantasy and is happy to put time, thought, and energy into living out a romantic fantasy with you. This doesn't mean he's pretending. He, in all likelihood, does love you and wants to have wonderful experiences with you. The problem comes if his romantic notions aren't balanced with a dose of realism. If being in love is the most important thing to him, he may have trouble with the parts of a relationship that are unromantic—like bills, sickness, death, in-laws. In short, he may not have sufficient engine power for the steep climbs or good shock absorbers for the rougher spots on the road of your relationship. The comedian Henny Youngman, as a wise old man in his 90s, gave this advice to young comedians who came seeking it: Fall in love and stay in love. The Romantic, if he can follow that advice, will be great husband material. If he can't, this guy may drive out of your life as soon as the roses start to wilt.

Most brands of men, if they are willing to put in the effort, can do a good job of being romantic. Romance is an optional feature that can be installed in almost all models of men. What we are describing with the Romantic brand is a man whose main identity is as someone who behaves romantically. There's a certain immaturity in that this man gets impatient with the less sweet sides of life and of relationships. He wants hearts and flowers and is not skillful in the messiness of changing dirty oil filters.

There's not really too much more to say about the Romantic because, in truth, romance is the icing. It's not the cake. Too much sugar without any protein is a bad diet for both a relationship and your body. We all love the new car smell, but once it's gone, you want him to still love riding in the relationship. Make sure your candidate is interested in your engine and offers a good maintenance contract and isn't just enjoying the new car smell. If that's all he wants, let him go to the car wash for a squirt of it.

## THE WORKAHOLIC

Unlike the Addict, this brand guy—the Workaholic—is predictable. You can predict he'll be working—early, late, on vacations, on Sundays, on his birthday and yours. "Overdrive" is his normal gear. He doesn't go into "Neutral" and "Park" without a hard push from you because idling is rough with the Workaholic. This model is impressive in the distance he can cover without having to refuel, but if the refueling station is home and you, waiting for him to make a pit stop may leave you feeling as lonely as the attendant inside a pump-and-pay-outside gas station.

The manufacturer defect in this guy is that he has a program chip in his brain that continually sends him this message: "You're OK as long as you are producing." If he stops working for long, a silent alarm goes off that shouts "Bad Boy! You're no good!" It's a serious glitch in the programming. Many great men were Workaholics, but not many great husbands were. Until this brand of man learns that he's OK and lovable even when he's in idle, he'll have a hard time hanging out in your garage.

In this society, which reinforces the work ethic, the Workaholic isn't likely to do anything to recover unless he's forced. Usually he surrounds himself with other workaholics, and one-upmanship becomes "I work (as in suffer) more than you do." He often likes his work, which makes giving it up, like giving up any drug, more difficult. The Workaholic carries a heavy load, so much so that he can't carry any of yours if you ask him to take over some part of it.

Workaholics are running big time. They are running from old pain that chases them like a nightmare that you can't shake. Take Brian. His sister died when both of them were teenagers. Then Brian and his wife lost a baby girl when she was two. Today Brian can't stop running from the pain. What was to be a leisurely car drive with his wife, her idea of an opportunity for talking and spending some intimate time together, became for Brian a torturous time when he was trapped with his feelings. He responded with so much criticism and irritability that his wife swore, to Brian's relief, they would never do that again.

Like the Addict brand, the Workaholic's ratings go up when he's working a recovery program. It may not be a formal 12-step program, but if he knows he has a problem with letting work fill up too many of the spaces of his life and is committed to making room for other things—like you, then he has the potential to make passing grades or even better on the essential features.

Those are the basic brands. Now see how they rate as husband material in Chapter 6.

## Six

## RATING THE BRANDS OF HUSBANDS

IT'S REPORT CARD TIME! As you will see, no brand of man gets all As, and no one flunks in every category (although the Bully comes close). There are no perfect husbands. If that's what you are waiting or looking for, abandon the quest. You aren't going to find one. After all, we are talking about real people here. No real human being, male or female, is nor should be perfect in every way. On the other hand, everyone has some redeeming value. We have tried to be fair and honest in assessing both the strengths and weaknesses of each brand of man. Our reasons for rating each as we do are explained. You may disagree with the grade we give on some feature if your experience with a brand was different. We are, of course, rating broad categories, not individual Bobs, Toms, and Leonardos. If the guy you have your sights set on scores As on every feature, congratulations! Just use our rating scale to be forewarned of potential problems that may arise when the relationship has more mileage.

The model you choose may fit more than one of our brands of men. For example, the Mercedes-Benz ML320 is a cross between a van and a traditional off-roader but handles like a car. It has a powerful engine but is also big enough to comfortably ride a family. (Yep, it's the same one that flunked the "moose test."). Like this car, your candidate may fall into more than one category. Common combinations are Eagle Scout/Power Seeker/Outdoorsman (rates high) and Macho Man-Narcissist-Power Seeker (rates *very* low). He could be a Mama's Boy/Romantic/Addict or a Jock/Peter Pan/Bubba. If your guy seems to be a multi-category model, look at the report card for each brand you think he fits in. That will give you a composite picture of what to expect.

Our Rating system is based on consumers' frequency of repair and reported satisfaction of ownership and performance. Here's what the grades mean:

*A*—means excellent performance. Demonstrates exceptional effort and skill in meeting the requirements.
*B*—means above-average performance. Delivers reasonable performance and a good ride.
*C*—means commendable performance. Applies sufficient effort to make a passing grade but nothing to surpass the ordinary expectations.
*D*—means passing but below-average in effort and performance. Has just enough effort and skill not to have to repeat the course.
*F*—means deficient and not acceptable in meeting the requirements.

And now (drum roll, please!), *The Consumer's Guide to Husband Material* Rating!

| BRAND | Commitment | Communication | Faithfulness | Friendship | Honesty |
|---|---|---|---|---|---|
| Addict | B | F | F | C | F |
| Big Daddy | A | C | A | B | B |
| Bubba | B | C | B | B | B |
| BULLY | A | F | C | D | F |
| Couch Potato | A | B | A | C | A |
| Eagle Scout | A | B | A | A | A |
| JOCK | C | D | C | C | B |
| MACHO MAN | C | B | D | B | C |
| Mama's Boy | F/A | B | A | A | B |
| Narcissist | F | C | F | D | A/F |
| Outdoorsman | A | C | A | B | B |
| Peter Pan | F | B | C | A | A |
| Poet | A | A | B | A | A |
| POWER SEEKER | A/F | B | B | A | B |
| Romantic | F | A | F | B | B |
| Workaholic | F | D | D | C | B |

# RATING THE BRANDS OF HUSBANDS

| Humor | Intimacy | Maturity | Reliability | Responsibility | Romance | Sex |
|---|---|---|---|---|---|---|
| B | D | F | F | C | B | A/F |
| C | C | A | A | A | A | B |
| A | D | C | C | B | C | A |
| D | F | F | F | C | F | C |
| B | D | C | B | C | D | C |
| B | B | A | A | A | B | C |
| C | D | D | C | C | B | A |
| A | C | B | B | B | A | B |
| B | B | D/A | A | A | B | C |
| A | F | D | F | B | A | B |
| B | B | B | A | A | C | B |
| A | A | F | F | D | D | C |
| C | B | C | C | C | B | C |
| B | C | A | B | A | A | C |
| B | A | D | F | C | A | A |
| C | D | B | C | A/F | C | B |

## THE ADDICT

A qualifier must be inserted for the scores given the Addict brand of man: All grades are likely to be higher if he is actively in recovery. A man who knows he has a problem with addictions and is taking serious steps to overcome it is doing the necessary repairs and maintenance we have talked about in earlier chapters.

*Commitment:* **B**—Addicts are basically dependent. They don't leave relationships easily as long as they can feed their addictions and still stay in the relationship. They end up getting left when a partner is no longer a co-addict or when it becomes too hard to live with their addictive behaviors.

*Communication:* **F**—Unless he is actively in a recovery program, he is going to fail in communication. Addictions serve the purpose of blocking emotions, so the nonrecovering addict is out of touch with his feelings, especially painful and vulnerable ones. He doesn't talk about his feelings because he doesn't feel them. He's not going to hear you when you talk about your feelings either.

***Faithfulness:* F**—His first loyalty (if he's not in recovery) is to his addiction, not you. You may matter to him very much, but his love for you will not deter him from lusting or lurching after his drug of choice, be it sex, booze, drugs, food.

***Friendship:* C**—The addict can be great fun and really be there for you at times, even a lot of the time, but his addiction is the only true friend.

***Honesty:* F**—Addictions make liars out of everyone sooner or later. The shame that drives addictive behavior forces honesty into hiding.

***Humor:* B**—Some of the greatest comedians were (and are) serious addicts. Some of them are also now dead. Addictions are not funny, even if the Addict sometimes is very much so.

***Intimacy:* D**—With an Addict, intimacy is sporadic and unpredictable, depending on the severity of the addiction. What can really complicate the picture is that this brand of man may be most intimate when he is acting out his addiction, such as when he's drinking or caught up in the passion of a sex addiction.

***Maturity: F***—He fails on maturity because maturity means he'll make the choice to do the grown-up thing when such a choice has to be made. With an Addict, you can't know that's the choice he'll make.

***Reliability: F***—A reliable Addict is an oxymoron. He isn't likely to be both. Addictions are inherently unpredictable. You don't know that you can count on him because you never know, with confidence, when he will choose his addiction over you and your needs.

***Responsibility: C***—Addicts can be responsible. They can hold good jobs and be successful in many aspects of life. But he gets a "C" because he is not responsible in the areas of his life where his addiction controls his decisions.

***Romance: B***—Addictions can either augment or destroy romance. Days of wine and roses can turn into mornings of hangovers and shame.

***Sex: A/F***—Sex, if it is his addiction, may be great, but it probably won't be great just with you. Excessive alcohol or drugs are certainly going to interfere with performance, even if they initially aid in getting it going.

## THE BIG DADDY

The scores given here are presuming the man fits mainly in this category. There is some tendency for the Big Daddy model to also be a Narcissist and/or a Power Seeker. If his Big Daddy role is combined with either of those brands, the scores can be very different.

*Commitment:* **A**—This is a man who wants to be there for you. He wants and needs you to need him.

*Communication:* **C**—The relationship, unless it matures, has a parent-child quality. Communication as equals is hindered by this covert understanding between both of you.

*Faithfulness:* **A**—The Big Daddy wants you to stay around because psychologically he needs you as much as you think you need him.

*Friendship:* **B**—He'll try to please you and get you what you want as long as you are a good girl. A parent enjoys having his kid around as long as she's fun to be with and reasonably well-behaved.

*Honesty:* **B**—You may not know everything important that's going on in his life. It's not so much a matter of deception as his belief that he doesn't need "to worry your pretty little head."

***Humor: C***—A Big Daddy may be inclined to take himself and his role in your life a little too seriously to be able to step back sufficiently to laugh at himself and see the humor in life.

***Intimacy: C***—You may see his vulnerabilities and weaknesses, but it won't be because he confided in you about them. True intimacy is limited by ego and the roles we play. To the extent you and your Big Daddy are playing out roles, intimacy between you will be comparably limited.

***Maturity: A***—The Big Daddy doesn't have a problem with being mature, but he may have a problem with your wanting to be the grownup, too.

***Reliability: A***—He's likely to be "a man of his word" because he wants you to rely on him.

***Responsibility: A***—He's used to taking responsibility for his own and others' lives, so he's not going to have a problem with carrying more than his share of the load as long as it fits with his role expectations.

***Romance: A***—Having a romantic relationship with "his girl" is part of the Big Daddy's fantasy of what the relationship is all about.

***Sex B***—Sex will be more than adequate as long as there's nothing you need that threatens him or unless he's having problems of his own that interfere with his performance.

## THE BUBBA

***Commitment: B***—The Bubba wants to belong to someone, so he's not likely to go wandering off, and he would be devastated if you decided to leave him.

***Communication: C***—He's not a deep thinker about feelings and hasn't mastered the fine details of communication, but he'll listen and try to understand once you get his attention. Don't expect a lot of self-disclosure from the Bubba.

***Faithfulness: B***—Kids get distracted. The Bubba's attention may wander not from lack of caring but because something or someone else caught his interest for the moment. A certain tolerance and patience for bringing him back on task is required in a relationship with a Bubba.

***Friendship: B***—He'll think of you as his best friend even if he forgets all about you from time to time.

***Honesty: B***—Bragging or exaggerating may be a problem with the Bubba, but he's not likely to be a skillful liar. He's probably too honest to seriously cheat on his income tax, even if he resists filing it until the last nanosecond.

*Humor:* **A**—Kids are cute, and so is the Bubba in many ways.

*Intimacy:* **D**—Because he doesn't think much about what he feels, he's not particularly skillful at expressing feelings nor listening to you discuss yours.

*Maturity:* **C**—The Bubba is never going to be an astronaut. He doesn't take life seriously enough to be one and isn't all that interested in being mature, if truth be told.

*Reliability:* **C**—You know you can't always count on kids to do what you need them to do, and the same applies to this brand of man.

*Responsibility:* **B**—He may procrastinate about getting done what he has to do, but most of the time the Bubba will carry his share of the load.

*Romance:* **B**—For your anniversary, you might get diamonds or a bread-making machine. It depends on what he imagines would be the perfect gift for you. He wants to please you but may not ask how best to do that.

*Sex: A*—He gets high marks here because sex is important to the Bubba. It is an important symbol to him that he's loved and wanted. He may require some instruction on how to be a sensitive lover, though.

## THE BULLY

Every class has its dunce, and in the school of husband material the Bully is it. Even so, he gets high marks and passing grades on some features.

*Commitment: A*—The relationship matters a great deal to this kind of man, too much so, in fact. He is going to be emotionally stuck to you like the glue that accidentally sticks your thumb and fingers together.

*Communication: F*—The Bully won't hear what you need because he's only tuned in to his own needs. Any complaints from you will be perceived as direct or implied criticism, which he will respond to by attacking.

***Faithfulness: C***—He's not necessarily going to run around, but he could if he feels entitled to for whatever he thinks is a valid reason.

***Friendship: D***—At the beginning, his excessive attentiveness may seem like he's your best friend. It's an attentiveness based on control, not friendship.

***Honesty: F***—The Bully is incapable of being honest with himself or anyone else. His whole psychological agenda is about being on top. Honesty is irrelevant.

***Humor: D***—It's hard to laugh with someone who can't laugh at himself and is so hypersensitive about how others, including you, perceive him.

***Intimacy: F***—Intimate moments with the Bully are like the calm in the eye of the hurricane. It's going to be dangerous going in and coming out of that intimacy.

***Maturity: F***—Bullies lack emotional intelligence. They haven't learned to override and control their emotional responses.

***Reliability: F***—Since reliability means you can count on him, the Bully will fail you big time on this feature.

***Responsibility:* C**—You may be able to count on him to go to work, bring home a paycheck, and show up on time. What happens once he shows up is up for grabs.

***Romance:* F**—The courtship may be impressive (see Friendship above), but intimidation and violence have never been considered love potions.

***Sex:* C**—Sex is likely to be troublesome, but it will happen. If the bullying extends to sex, it's likely you would rate this aspect of the relationship much lower.

## THE COUCH POTATO

***Commitment:* A**—Your Couch Potato guy isn't going anywhere. Wandering takes energy, and he's not going to expend it looking for another relationship unless he's really miserable.

***Communication:* B**—Communication takes energy, too. He can do it, though, once you get his attention.

***Faithfulness:* A**—He doesn't have any reason not to be faithful, and his basic complacency works in favor of it.

*Friendship: C*—His idea of friendship, having you sit by him on the couch in front of the TV, gets just a passing grade.

*Honesty: A*—He doesn't lie because he doesn't do anything that requires a cover-up. Besides, lying takes energy.

*Humor: B*—The Couch Potato is often a good-natured sort of guy with as much readiness as anyone else for having a laugh. Because he doesn't take himself terribly seriously, he doesn't mind if you don't either.

*Intimacy: D*—Boob-tubing can be a defense against closeness, a way to hide from you and from himself. It's hard to feel intimate with a man who's channel surfing.

*Maturity: C*—Maturity entails, in part, putting goals above wants. The Couch Potato guy has a tendency to put life on hold. "Wait until the commercial comes on!" is his basic existential position.

*Reliability: B*—Since the Couch Potato is disinclined to make unnecessary changes, he's predictable. You'll know where he is, and once you've got his attention, he can be there for you.

***Responsibility: C***—He'll get the chores done, with some persistent nagging. Eventually he'll have sufficient guilt about not doing what he needs to do that he'll get on with it.

***Romance: D***—Romance takes work, so don't expect much.

***Sex: C***—It may have to be squeezed in during the commercials, so don't expect it to be great. Try to tune into infomercials if you want some foreplay, and plan on celibacy during the Super Bowl. He won't even want to miss the commercials then.

## THE EAGLE SCOUT

***Commitment: A***—Remember the Scout Law? The second quality of a scout is loyalty. This brand takes promises and vows seriously.

***Communication: B***—He may have trouble talking about unpleasant things that he thinks might hurt you, things that really do need to be discussed sooner or later.

***Faithfulness: A***—This gets back to loyalty again but also includes "trustworthy," the first requirement of the Scout Law.

*Friendship: A*—He's been part of a pack since he was a kid and has learned the art of being a good friend.

*Honesty: A*—See "trustworthy" under the Scout Law.

*Humor: B*—He takes life and his role in it seriously, but that doesn't mean he can't see the lighter side of life.

*Intimacy: B*—He'd make an A if he weren't so worried about saying and doing the right thing.

*Maturity: A*—What else would you expect from someone who got all those merit badges?

*Reliability: A*—It's not part of the Scout Law, but you can count on this brand guy to come through in a predictable way.

*Responsibility: A*—This is Mr. Responsible. Peter Pan guys would do well to apprentice with the Eagle Scout model.

*Romance: B*—He'll do a good job of being romantic on appropriate occasions, but don't expect him to pull off the highway to have an unplanned interlude in the bushes.

*Sex: C*—See Romance above.

## THE JOCK

***Commitment: C***—His first commitment is to his pecs and his image.

***Communication: D***—He's not likely to be interested in hearing what you want and need emotionally.

***Faithfulness: C***—His developmental age is about sixteen. Teenage relationships are not famed for their faithfulness.

***Friendship: C***—If you're in the gym with him, you're his friend. He doesn't really get it about how to have friends outside of that sphere.

***Honesty: B***—Unlike the Eagle Scout, he really is going to be honest. It won't occur to him to ponder whether or not what he's going to say will hurt your feelings.

***Humor: C***—Adolescents take themselves very seriously.

***Intimacy: D***—Being vulnerable and open as is required in intimacy is really too scary for the Jock.

***Maturity: D***—Do we need to say more?

***Reliability: C***—He shows up when it serves his own interest. Don't count on him to be there if you suddenly feel faint and start to fall backwards.

***Responsibility: C***—He'll make sure he gets there in time to open the gym in the morning, but getting him to fix broken equipment or adjust it to your needs is another story.

***Romance: B***—He'll be sure to do enough to tell his workout buddies all the things he does for his wife. He'll put on a good show for the occasion.

***Sex: A***—Performance is important to this model. He'll work at it like he works on his physique.

## THE MACHO MAN

***Commitment: C***—This model is always thinking about upgrading.

***Communication: B***—It won't occur to him not to tell you what he's thinking and feeling (unless he thinks you'll get pissed off, which you might). He'll hear what you have to say. He just might not think it's necessary to do anything about it.

***Faithfulness: D***—See Commitment above.

***Friendship: B***—He can be great fun to hang out with as long as he's having a good time.

***Honesty: C***—This brand is his own best "spin doctor."

***Humor: A***—He is a performer and the best are good at it.

***Intimacy:* C**—Intimacy involves being real, which comes and goes with the Macho Man.

***Maturity:* B**—He usually can pull off things in the world (as in being President), but, like a little kid, he has trouble resisting his impulses, especially those emanating from the crotch.

***Reliability:* B**—For the most part, you can count on him.

***Responsibility:* B**—He is willing to take responsibility unless it's for something he does wrong.

***Romance: A***—It's part of his image.

***Sex: B***—Ditto Romance above. He loses a point because his performance is likely to be on his schedule rather than your mutual schedules.

## THE MAMA'S BOY

***Commitment:* F/A**—He's not yours as long as his heart belongs to Mama. Once it belongs to you, he's yours forever.

***Communication:* B**—He'll hold back from telling you what he thinks you may not want to hear, and he has a tendency to hear your upsets as criticism of him.

***Faithfulness: A***—It's part of being a good boy, plus he doesn't want to leave you.

***Friendship: A***—This brand works at being a thoughtful friend, and he likes women.

***Honesty: B***—It's not nice to lie (but see Communication above).

***Humor: B***—He has a little trouble lightening up if he thinks he's done something wrong.

***Intimacy: B***—He's open to his feelings and yours but can withdraw into feeling hurt at times.

***Maturity: D/A***—He gets top grades, once he's broken free of Mama.

***Reliability: A***—It's the stuff good boys are made of.

***Responsibility: A***—See Reliability above.

***Romance: B***—He'll put a lot of thought into it, but it's likely to be planned passion.

***Sex: C***—You can't get too wild if you're thinking about behaving well.

## THE NARCISSIST

***Commitment: F***—If he can't have his way, you're history.

***Communication: C***—It's sporadic. He's best at telling you what he wants and how mad he is that you didn't give it to him. Anything you say you want gets heard as unfair criticism.

***Faithfulness: F***—Like the Macho Man, the Narcissist is keen on upgrading.

***Friendship: D***—Some of us have friends like this—for a while.

***Honesty: A/F***—He can and will be brutally honest about what he feels. He also can lie much better than Pinocchio.

***Humor: A***—He's funny—for a while—but never has a sense of humor about himself.

***Intimacy: F***—You can't be close to someone who's that closed off from his real self.

***Maturity: D***—Two-year-olds get the same grade.

***Reliability: F***—The only thing he's consistently reliable about is his emotional unpredictability.

***Responsibility: B***—Many Narcissists have good jobs, the better to buy you with, my dear.

***Romance: A***—He puts on a great show.

***Sex: B***—Sex is how he reassures himself he's loved. It has little to do with loving you.

## THE OUTDOORSMAN

*Commitment:* **A**—It's part of the traditions he believes in. He's not going to leave you any more than he's going to give up his duck lease.

*Communication:* **C**—The grade is higher if the subject is a fishing/hunting trip.

*Faithfulness:* **A**—See Commitment above.

*Friendship:* **C**—He knows how to be a friend and expects to be one with you.

*Honesty:* **B**—It's part of his character to tell the truth, unless it's about the one that got away.

*Humor:* **B**—The Outdoorsman enjoys life and sees the fun in it.

*Intimacy:* **B**—He won't hold back in sharing his feelings or hearing yours, but he may not give much thought to either.

*Maturity:* **B**—He'll do his homework and chores, but he's always ready to slip out the back door to play.

*Reliability:* **A**—To be competent as an Outdoorsman, he has to be reliable.

*Responsibility:* **A**—He knows it's important to do what's required if disasters and accidents are to be avoided.

***Romance: C***—He won't see much reason to put extra effort in this area of life. Think candlelight in the fishing cabin.

***Sex: B***—The Outdoorsman is a natural man, as in primitive. He's OK with doing what comes naturally.

## THE PETER PAN

***Commitment: F***—Peter Pan guys have the commitment capacity of nine-year-old boys. They would love to come over to your house to play, until they get bored or get a better offer.

***Communication: B***—He is sufficiently guileless to be straightforward about what he's thinking and feeling, and he'll believe you when you tell him what's going on with you, even if he's baffled as to why you'd think or feel as you do.

***Faithfulness: C***—He's not going to betray you. He just may forget to think about you. (See Communication above.)

***Friendship: A***—Peter Pan is a great friend. He delights in showing you the adventures of life and will be there when Captain Hook attacks.

***Honesty: A***—What's there to hide? This is not a tricky guy.

*Humor: A*—Life is fun and funny for Peter Pan.

*Intimacy: A*—Like a child, the Peter Pan guy is open with himself and to you.

*Maturity: F*—Remember his favorite song? *I Don't Won't Grow Up.*

*Reliability: F*—Don't count on this guy to take out the trash without being reminded.

*Responsibility: D*—He tries to do what he's supposed to do, but, darn it, something else more interesting keeps coming up.

*Romance: D*—Ever had a date with a nine-year-old boy?

*Sex: C*—It's hard to sustain a passionate relationship when you feel like you're his mother.

## THE POET

*Commitment: A*—Relationships may be hard for the poet, but he sticks by what and whom he believes in.

*Communication: A*—Expressing himself and understanding what others are saying is his passion.

*Faithfulness: B*—If you are his soul mate, you become part of the poetry of his life.

*Friendship:* **A**—He wants to share his life's experiences with you and be a part of yours.

*Honesty:* **A**—Part of his métier is saying what is hard to say and hearing what's really being said.

*Humor:* **C**—Poets need someone to help them not take themselves too seriously. The Poet Laureate Richard Wilbur, in commenting on a poem, said his wife, in hearing the poem, responded, "Congratulations! You have finally written a poem that makes absolutely no sense at all."

*Intimacy:* **C**—His grade comes down a little here because the Poet sometimes has trouble letting you know what's going on with him until he's ready to express it in his way. By the same token, if he's deep in his inner work, he may not be available to what you need to express.

*Maturity:* **C**—It takes a certain maturity to be comfortable with the mundane of the world, like saving receipts for tax deductions. The Poet may forget that these things count, too.

*Reliability:* **C**—The Poet brand of man needs more inward time than some other models, time when he recharges his psychic batteries. You have to plan for his not being available to be ready to go on the road sometimes and needing more time in the garage than you might prefer.

***Responsibility: C***—(See Reliability above.) He wants to help, but sometimes he gets lost in his own intellectual/emotional world.

***Romance: B***—Romance is not an ego thing for the Poet. He doesn't pride himself on being a Don Juan. Romantic interest can fade quickly if his mood turns inward, but returns just as quickly after the dark clouds of introspection have passed. You have to be patient with the Poet.

***Sex: C***—The Poet guy's passion is from his head and heart more than in his crotch. He can perform perfectly adequately and will want to satisfy you, but he probably won't spend a lot of time thinking about sex except when the occasion (among other things) arises.

## THE POWER SEEKER

***Commitment: A/F***—The Power Seeker sees a relationship like a company he owns. He expects to work at it and be involved in it. If he decides it's a losing proposition, however, he'll cut his losses (which means you) and move on.

**Communication: B**—He tells it as he sees it, but he may have some trouble hearing it from your perspective.

**Faithfulness: B**—The Power Seeker wants to win and that includes making his relationship with you successful.

**Friendship: A**—He needs and wants a supportive woman by his side to be his friend.

**Honesty: B**—This brand man isn't afraid to tell the truth. He just may not take time to think about what the truth is for him.

**Humor: B**—The Power Seeker sees the funny side of life.

**Intimacy: C**—He can get close to you, but he may have some real trouble when it comes to expressing vulnerability. The Power Seeker isn't embarrassed to cry with tears of joy or gratitude, but he's not likely to be open about fear—his or yours.

**Maturity: A**—It takes maturity to have "the right stuff." He does what it takes to get the job or the deal done.

***Reliability: B***—If the Power Seeker says he's going to do something for you, he'll do his best to do it.

***Responsibility: A***—He carries heavy loads with good shock absorbers. Just don't expect him to notice which loads you need carrying. The motto with your Power Seeker, with regard to his helping you out, needs to be "Ask and ye shall receive." It's not "The meek shall inherit the earth."

***Romance: A***—The Power Seeker likes to think of himself as successful in everything he undertakes. That includes being a good lover. He may not do it your way, but he'll try to do what he thinks is the right way.

***Sex: C***—Sex is only one area of potency for this brand man. He gets the deal done and moves on.

## THE ROMANTIC

***Commitment: F***—Love is a mood thing for the Romantic brand. Moods change.

***Communication: A***—He loves to tell you what's in his mind and heart and know what's in yours.

***Faithfulness: F***—See Commitment above.

*Friendship: B*—He's great for long talks, walks, and drives in the country. But a friend-who'll-be-there-no-matter-what he isn't.

*Honesty: B*—The Romantic brand may hold back from disclosing what's not so romantic about himself or your relationship.

*Humor: B*—He gets graded off for his idea of how life ought to be taken too seriously.

*Intimacy: A*—The Romantic runs on high octane intimacy.

*Maturity: D*—As romance fades, so does he.

*Reliability: F*—He's not going to fail to show up with a picnic basket for your date, but he may fail to tell you that he's falling out of love with you.

*Responsibility: C*—The Romantic doesn't really want responsibility for making sure the routine maintenance is done on the relationship so that it runs well for those long drives in the countryside.

*Romance: A*—He's an Indy 500 winner on this track.

*Sex: A*—It's part of the image and the fantasy that keeps his engine purring.

## THE WORKAHOLIC

***Commitment: D***—His commitment is to his work.

***Communication: D***—Why do you think he works all the time? It's a great escape from having to talk about messy things, like feelings, needs, your relationship with each other, stuff like that.

***Faithfulness: D***—He's faithful to working. Some people find safety and solace in prayer. This brand guy finds it in work.

***Friendship: C***—When he's not on the work highway, he can be as good a friend as the next guy. He's just not one you can expect to spend much time hanging out with.

***Honesty: B***—He doesn't lie, but he uses work as a justification for not doing what he doesn't want to do.

***Humor: C***—The Workaholic takes himself and his responsibilities too seriously. He forgets the bigger picture.

***Intimacy: D***—The work blocks closeness. That's the drug of choice for the Workaholic. He can't share feelings he can't feel—yours or his.

***Maturity: B***—This guy tries hard to do the right thing. His immaturity is in having to perform all the time.

***Reliability: C***—You can't count on him to show up for extracurricular activities of your life together.

***Responsibility: A/F***—The "A" is for work. The "F" is for you.

***Romance: C***—Out of guilt, he'll sporadically pull together a romantic gesture. There was a credit card ad that showed a guy caught forgetting his anniversary and covering by making up, on the spot, the surprise trip he had planned for her to some exotic land. To save face with her, he had to carry through. Maybe you'll get lucky like that with your Workaholic. Just don't count on it.

***Sex: B***—The performance will be adequate or better when it happens. Whether that happens enough to suit you depends on your maintenance schedule.

There you have the ratings. You have to look over the report card your guy gets and decide for yourself if he's suitable husband material for you. If he is and you've decided to choose him as *the one*, the last chapter, Living With Your Choice, will give you some ideas on how to avoid buyer's regret.

## Seven

## LIVING WITH YOUR CHOICE

WOMEN SHOP DIFFERENT WAYS. Some want to know all the options. They go from store to store, checking prices and availability. After they have gathered the information and made comparisons, they decide on their selection. Other women use the fit-the-criterion approach. If it's what they are looking for and the price is right, they cease the search and get it. Finally, there are the impulse buyers. They like what they see and take it. How you shop for clothes, shoes, and cars may color how you shop for husband material. We hope using our *Guide* will help you curb any tendencies to impulse buying at least long enough to be an informed consumer. But whatever your approach, once you have found and chosen your brand man, you have to live with your choice.

Buyer's regret is a normal psychological phenomenon. We always wonder, if fleetingly, whether we made the right choice. "Maybe I could have gotten a better deal." "Maybe this is not the right one after all." "Maybe I should have looked longer." Those "maybes" kick in disturbingly quickly. The trick comes in knowing what to do with them once we start hearing them in our heads. The purpose of this chapter is to give you some ways of thinking about those "maybes" and about your choice of husband that can help quiet the voices.

Here are five useful lessons to remember once you have made your choice:

1. No man is perfect, as in flawless, but he may be perfect for you.
2. Learning to live with and love another human being as a mate is a lifelong process.
3. Every one who is important to us, husbands included, is a valuable and essential teacher for what we need to learn in life.
4. The man we choose is a mirror of the masculine side of our own personalities.

5. Life is too short to spend it regretting your choice.

## LESSON #1:
*No man is perfect, as in flawless, but he may be perfect for you.*

Definitions for perfection are "being entirely without fault or defect: flawless" and "corresponding to an ideal standard." Forget those when it comes to husbands. It's not going to happen. But a more useful definition of perfection can apply: "satisfying all requirements." In Chapter 4, we had you fill out questionnaires that helped you understand what your criteria are for a husband. Take seriously what you wrote and concluded. The man who comes closest to meeting those criteria is the right man for you. It doesn't matter that he does things that drive you nuts (as long as those things aren't destructive to you or himself). That he falls far short of an idealized image, yours or one you see projected on the big screen or magazine covers, is completely irrelevant. If he makes passing grades on the essential features of husband material and you want to love and live with him, don't indulge in buyer's regret. Put your energy in lesson #2.

## LESSON #2:
*Learning to live with and love another human being as a mate is a lifelong process.*

Human beings are much more complicated than automobiles. Once you have figured out how to drive your Toyota or Geo or Mazda or Dodge, you've pretty much got it. True, as it gets old and develops idiosyncrasies, you have to learn to adjust to its quirks. But nothing you will ever have to learn about a car compares with what you will have to learn about a husband—or what he will have to learn about you. It's a lifelong learning class and not necessarily a leisure one either. The difficulty in learning about human beings is that they keep changing. Cars don't go through mid-life crises. They don't mourn or get depressed or have broken dreams. Even Jaguars or Mercedes don't have inflated egos. The drivers may, but the autos don't. Cars don't have addictions or unresolved mother complexes. They don't worry that their engines are too small or have opinions about what brand gas you fill them with. But you aren't looking to take a car as your husband, even if you hope to keep your car for the rest of your life and consider it a great friend. People fall in love with their cars, but they don't marry them.

So, once you take a husband, remember this: **you do not now nor will you ever know him completely.** The longer you live as husband and wife, the more you will know about him, but the course of study never ends. He will change, hopefully growing in wisdom, grace, and maturity, and you will change, ideally in the same direction. Unless you cling to a rigid perspective, the way you see him now is not identical to the way you saw him when you first met nor the way you will see him in the future. It can't be the same because no human stays exactly the same. For better or worse, we keep changing. Keeping this in mind, allow yourself to always look at your choice of man with fresh eyes. In Zen teaching, they call it "beginner's mind." Look at him each day as if you had never seen him before. That way you will always be open to learning who he really is. The trouble comes when we get fixed in our view of another person. It doesn't allow them to change in our perception. Living with another human being is the most challenging task a person can undertake, but we must be up to the challenge since we all start learning how to do it even before we're born. So now that you have made your choice of husband, welcome the opportunity to discover all you can in a lifetime with this man.

## LESSON #3:
***Everyone who is important to us, husbands included, is a valuable and essential teacher for what we need to learn in life.***

We told you stories in this book about what we learned from cars. From her 1950 MG-TD, Rita learned about the need for balancing play and practicality. From her Mercedes, Kimberly learned about the importance of deciding for herself what she needed to take her through life. Those are mundane, if useful, lessons compared to what any woman learns from her choice of husband. We don't believe there are wrong choices. There are just some teachers that are very tough and some lessons that are very difficult to learn.

There's an old saying, "When the student is ready, the teacher appears." A husband is a unique teacher in any woman's life. He will teach you as much about yourself as you will learn from anyone—mother, father, children included. Our belief is that what he teaches you, regardless of how the course is taught, is exactly what you need to learn at that moment in your life. When we choose our husbands, we choose our teachers. If what you need to learn is how to love someone, he'll teach you that. If you need to learn how to stand up for yourself, he will teach you that. The instruction may not be direct or kindly, but he will teach you nonetheless.

Let's look at some examples. Sharon's father abandoned the family when she was eight. Her mother, left with four young daughters, fell into a deep depression, from which she emerged only after the daughters were grown. Sharon had no role model for how a woman takes charge of her life and responsibility for the lives of her children. When she married, Sharon chose Tom, an Addict brand man. Tom was an alcoholic, as her father had been, and neglected her and their young son. Eventually, Sharon decided the situation was intolerable. Facing enormous terror at the idea of being alone and on her own, Sharon divorced Tom. What that husband taught Sharon is that she could indeed take charge of her life and take action for the benefit of herself and her child—something her mother had never been able to do.

Happily, it's not necessary to have a bad husband to have a useful teacher. Catherine had a domineering mother, a matriarch who ran her family with the precision of a military general. Catherine chose David as her husband. David is a combination of Power Seeker and Mama's Boy. He was loyal to Catherine, as he had been to his mother, but he was also a man who was confident in his goals and determined to reach them. He taught Catherine how to curb her need for control and to respect his autonomy. She wasn't a willing student, but, because he was loyal and committed, he persisted, and they grew in respect and appreciation for each other's strength. She learned the lesson she needed to learn: how to take charge and still be appreciative of the needs of others. The lesson was invaluable both in her relationship with David and their children and in her career.

The point is that it is no accident you choose the brand man you do as husband. Like it or not, you are signing up with precisely the right teacher for what you need to learn in your life's journey. After all, that's what life is supposed to be—a journey to enlightenment, wisdom, and love. Who better to help us learn than a husband?

## LESSON #4:
*The man we choose is a mirror of the masculine side of our own personality.*

The guy you choose is a part of you. Specifically, he's a reflection of your masculine side. It doesn't matter how different you think he is from you. Man or woman, we all have both masculine and feminine sides to our personality. The masculine side is the part of us that pushes out into the world, makes discriminating judgments, eliminates what's unnecessary, and is creative. Think of it as being the penis part of your personality. The feminine side is concerned with maintaining and building relationships, holding things together, and sustaining action. It's the womb part of the personality.

Some women, like Catherine's mother whom we just described, have a strong, even tough, masculine side. Others, like Sharon's mother, have a weak masculine side that is almost nonfunctional. Mentally healthy people have a balance between their masculine and feminine sides, as in men who can cry and women who can be strong. We get off balance when we use either the masculine or feminine side too much. It's like never having your tires rotated. When the balance is off, your psychological ride is rough. The reason we are telling you all this is that it's helpful in understanding your choice to consider that the man you pick is, in all probability, a manifestation of your own masculine side.

It may be a part of you that is unexpressed or even unknown. Think of it as being optional features on your car or computer that you don't use because you didn't even know you had them.

This bit of information can be either encouraging or upsetting. It's encouraging if the man you chose is honest, clear, and thoughtful. So is your masculine side. If, on the other hand, your man is boorish and inconsiderate of your feelings, it means you would do well to look at how your own internal masculine treats yourself. Are you inconsiderate of your own feelings? You may be attentive to your man's needs (and everyone else's), which is a measure of a well-functioning feminine side, but do you treat yourself boorishly? If you are attracted to the Bully brand, consider how you bully yourself. This is not an easy question to ask the mirror-mirror-on-the-wall. So most women don't. They just blame the guy for being like he is, which is OK, but it doesn't get you very far in living with your choice.

## LESSON #5:
*Life is too short to spend it regretting your choice.*

Whether you are a first-time consumer of husband material or more experienced than you wish you were in selecting husbands, there's one piece of advice we'd like to leave with you. Until you know for certain that you have chosen wrongly, don't look back. There is no point in wondering if you made the right choice. The choice you made is the choice you made. Start living in the relationship as if you had made the best choice in the world. For in truth, given all the variables, that's what you did. No one is omniscient. Neither you nor your brand man knows how this is going to turn out. Every choice has consequences. Part of being a mature consumer is being willing to accept the consequences of your choice. We are not talking about "You made your bed. Now lie in it." This isn't advice to stick it out no matter what. What we are suggesting is that you not spend time looking back over your shoulder at the brands left on the lot. That does you and your man a disservice, and no marriage can thrive without a true commitment to the marriage vow of "forsaking all others." So pat yourself on the back for making the wisest choice you can and move on into making a life with him.

We said this advice is conditional: "until you know for certain that you have chosen wrongly." Wrong has many definitions: not correct, erroneous; contrary to conscience, morality, or law; wicked; unfair or unjust; not required, intended, or wanted; not fitting or suitable; inappropriate; improper; not functioning properly; out of order; amiss. (*The American Dictionary of the English Language,* William Morris, Ed., Houghton Mifflin Co, Boston, 1981, p.1478)

No one can decide for you when your choice of husband fits any or all of those criteria. We urge you to give yourself time to think and seek professional advice before coming to that conclusion. Remember, this was supposed to be "for better or worse." But once you have done sufficient soul-searching and an in-depth psychological inventory on yourself, if your choice is to let go, then do so without regret. If you must have regrets, have them quickly and move on. It's good to grieve for awhile, for cars and men that are no longer in our lives. Just don't let the grief turn into self-flagellation. Remind yourself that, given everything, you made the best choice you could. You learned what you needed to learn (see Lesson #3 above), and it turned out as it was wont to do. Think of Frank Sinatra singing "I Did It My Way." That's the only way you can do it. Besides, there are still plenty of brands out there hoping you'll take them for a test drive and make them your own.

*INDEX*

Abzug, Bella 13, 26
Addict 92–94
   and recovery 142
   rating 140–141, 142–144
aggression 100–102
bad husbands
   in love with an image
      44–47
   incapable of caring 36–39
   physically abusive 43–44
   psychologically abusive
      40–42
   the wrong man for you
      47–50
   traits of 35–51
Big Daddy 94–96
   rating 140–141, 145–147
brands of husband material
   91–135
   rating 137–170
Bubba 96–98
   rating 140–141, 147–149
Bully 98–102
   and emotional intelligence
      150

   rating 140–141, 149–151
Chang, Valerie Nash 40–41,
   45
Cheney, Jane 116–117
choosing the right brand of
   husband 69–89
   want versus expectation
      74–77
   what am I looking for in
      a husband questionnaire
      79–85
Clinton, President William J.
   112–114
commitment 16, 19–20, 22,
   115–116
   and Addict 142
   and Big Daddy 145
   and Bubba 147
   and Bully 149
   and Couch Potato 151
   and Eagle Scout 153
   and Jock 154
   and Macho Man 156
   and Mama's Boy 157
   and Narcissist 158

   and Outdoorsman 160
   and Peter Pan 161
   and Poet 162
   and Power Seeker 164–165
   and Romantic 166
   and Workaholic 168
communication 16, 20–21, 22, 123
   and Addict 142
   and Big Daddy 145
   and Bubba 147
   and Bully 149
   and Couch Potato 151
   and Eagle Scout 153
   and Jock 155
   and Macho Man 156
   and Mama's Boy 157
   and Narcissist 158–159
   and Outdoorsman 160
   and Peter Pan 161
   and Poet 162
   and Power Seeker 165
   and Romantic 166
   and Workaholic 168
Couch Potato 102–105
   and depression 104
   rating 140–141, 151–153
depression
   and Couch Potato 104

Eagle Scout 105–109
   rating 140–141, 153–154
Einstein, Albert 18
emotional intelligence 19, 150
expectations
   a husband like your father 57–58, 74–76
   that don't match reality 59–64
fairness 28
faithfulness 16, 19, 21, 114
   and Addict 143
   and Big Daddy 145
   and Bubba 147
   and Bully 150
   and Couch Potato 151
   and Eagle Scout 153
   and Jock 155
   and Macho Man 156
   and Mama's Boy 157
   and Narcissist 159
   and Outdoorsman 160
   and Peter Pan 161
   and Poet 162
   and Power Seeker 165
   and Romantic 166
   and Workaholic 168
Forbes, Malcolm 130

# INDEX

friendship 16, 22–23
   and Addict 143
   and Big Daddy 145
   and Bubba 147
   and Bully 150
   and Couch Potato 152
   and Jock 155
   and Macho Man 156
   and Mama's Boy 158
   and Narcissist 159
   and Outdoorsman 160
   and Peter Pan 161
   and Poet 163
   and Power Seeker 165
   and Romantic 167
   and Workaholic 168
Goleman, Daniel 100–102
hearts
   and the capacity to love 26
honesty 16, 22, 23–24
   and Addict 143
   and Big Daddy 145–146
   and Bubba 147–148
   and Bully 150
   and Couch Potato 152
   and Eagle Scout 154
   and Jock 155
   and Macho Man 156
   and Mama's Boy 158
   and Narcissist 159
   and Outdoorsman 160
   and Peter Pan 161
   and Poet 163
   and Power Seeker 165
   and Romantic 167
   and Workaholic 168
Houston, Jean 66–67
humor 16, 24
   and Addict 143
   and Big Daddy 146
   and Bubba 148
   and Bully 150
   and Couch Potato 152
   and Eagle Scout 154
   and Jock 155
   and Macho Man 156
   and Mama's Boy 158
   and Narcissist 159
   and Outdoorsman 160
   and Peter Pan 162
   and Poet 163
   and Power Seeker 165
   and Romantic 167
   and Workaholic 168
husband material
   changing 5, 50–51
   traits of 14–33
intelligence 18–19
intimacy 16, 22, 25–27, 97

and Addict 143
and Big Daddy 146
and Bubba 148
and Bully 150
and Couch Potato 152
and Eagle Scout 154
and Jock 155
and Macho Man 156
and Mama's Boy 158
and Narcissist 159
and Outdoorsman 160
and Peter Pan 162
and Poet 163
and Power Seeker 165
and Romantic 167
and Workaholic 168
Jock 110–112
 rating 140–141, 155–156
Justice, Blair 26
Kanji 22
living with your choice
 171–184
love
 capacity to 25–27
Macho Man 112–114
 rating 140–141, 156–157
Mama's Boy 115–118
 rating 140–141, 157–158
marriage
 is it for you? 53–68

maturity 16, 27–28, 30,
 116, 123, 132
and Addict 144
and Big Daddy 146
and Bubba 148
and Bully 150
and Couch Potato 152
and Eagle Scout 154
and Jock 155
and Macho Man 157
and Mama's Boy 158
and Narcissist 159
and Outdoorsman 160
and Peter Pan 162
and Poet 163
and Power Seeker 165
and Romantic 167
and Workaholic 169
Miller, Lawrence E.
 118–119
Morrison, Jim 127
Narcissist 118–121
 rating 140–141, 158–159
obstacles to finding and
 selecting a husband 56–67
Outdoorsman 121–123
 rating 140–141, 160–161
Peter Pan 124–126
 rating 140–141, 161–162
 rescuing 124–125

# INDEX

physical abuse 43–44, 102
Poet 126–128
   rating 140–141, 162–164
Power Seeker 128–130
   rating 140–141, 164–166
psychological abuse 40–42
recovery
   and Addict 142
reliability 16, 28–29, 111, 128
   and Addict 144
   and Big Daddy 146
   and Bubba 148
   and Bully 150
   and Couch Potato 152
   and Eagle Scout 154
   and Jock 155
   and Macho Man 157
   and Mama's Boy 158
   and Narcissist 159
   and Outdoorsman 160
   and Peter Pan 162
   and Poet 163–164
   and Power Seeker 166
   and Romantic 167
   and Workaholic 169
rescue
   and Peter Pan 124–125
responsibility 16, 29–30, 125–126
   and Addict 144
   and Big Daddy 146
   and Bubba 148
   and Bully 151
   and Couch Potato 153
   and Eagle Scout 154
   and Jock 155
   and Macho Man 157
   and Mama's Boy 158
   and Narcissist 159
   and Outdoorsman 160
   and Peter Pan 162
   and Poet 164
   and Power Seeker 166
   and Romantic 167
   and Workaholic 169
romance 6, 16, 31–32, 123, 131–132
   and Addict 144
   and Big Daddy 146
   and Bubba 148–149
   and Bully 151
   and Couch Potato 153
   and Eagle Scout 154
   and Jock 156
   and Macho Man 157
   and Mama's Boy 158
   and Narcissist 159
   and Outdoorsman 161
   and Peter Pan 162

    and Poet 164
    and Power Seeker 166
    and Romantic 167
    and Workaholic 169
Romantic 130–132
    rating 140–141, 166–167
Scout Law 105–107
Scout Oath 105
sex 16, 32–33
    and Addict 144
    and Big Daddy 147
    and Bubba 149
    and Bully 151
    and Couch Potato 153
    and Eagle Scout 154
    and Jock 156
    and Macho Man 157

    and Mama's Boy 158
    and Narcissist 159
    and Outdoorsman 161
    and Peter Pan 162
    and Poet 164
    and Power Seeker 166
    and Romantic 167
    and Workaholic 169
Siegel, Bernie 35
Spock, Benjamin 116–117
Vaughn, Frances 49, 60–63
what-I-have-to-have-in-a-husband job description 86–89
Workaholic 133–135
    rating 140–141, 168–169

## RESOURCES FOR FURTHER RESEARCH ON HUSBAND MATERIAL

Harville Hendrix, Ph.D. *Getting the Love You Want: A Guide for Couples,* HarperPerennial, 1990.

Harville Hendrix, Ph.D. *Keeping the Love You Find: A Guide For Singles,* Pocket Books, 1993.

Valerie Nash Chang, *I Just Lost Myself: Psychological Abuse of Women in Marriage,* Praeger, Westport, Connecticut, 1996.

## ABOUT THE AUTHORS

—·—·—·—·—

RITA JUSTICE, Ph.D., is a clinical psychologist who has been in private practice in Houston for more than twenty-six years. With her husband, psychologist Blair Justice, she co-authored two pioneering books on child abuse and incest. Her most recent book, *Alive and Well: A Workbook for Recovering the Body*, addresses the body-mind connection. Rita has lectured and taught extensively internationally. She loves mountain climbing, her Tibetan Terrier Tashi, and travelling with her husband Blair.

KIMBERLY NELSON, D.J.M., is vice-president of marketing for GDH International–Alien Sport Division. Prior to joining GDH, she was principal owner of a sales and marketing company based in Houston. Kimberly loves playing drums, her nieces, and hanging out with her husband Jeff.

## ABOUT THE AUTHORS

Rita and Kimberly are available for lecturing.
Please contact them through:

Peak Press
2402 Westgate Drive, Suite 200
Houston, Texas 77019
Phone: 713-528-6680
Fax: 713-528-6577
e-mail: peakbook@flash.net

Rita and Kimberly aren't done with the subject of husband material yet. Two more titles are in the works: *Scoring As Husband Material: A Guy's Guide to Raising Your Ratings* and *The Consumer's Guide to Used Husbands*.

You are invited to contribute your funny story, favorite quotation, or bit of wisdom to these valuable new guides. Just send a copy of your story or contribution, indicating which book it's for, to the following address:

Husband Material (Specify which book)
c/o Peak Press
2402 Westgate Drive, Suite 200
Houston, TX 77019
Tel: 713-528-6680
Fax: 713-528-6577
e-mail: peakbook@flash.net

You will be sure to be credited for your submission.

Distributed by: BOOKWORLD
1933 Whitfield Park Loop
Sarasota, FL 34243
(800) 444-2524
Fax: (800) 777-2525
e-mail: sales@bookworld.com
www.bookworld.com